Mercenaries, Hybrid Armies and National Security

This book assesses the use of 'mercenaries' by states, and their integration into the national armed forces as part of a new hybridisation trend of contemporary armies.

Governments, especially in the West, are undertaking an unprecedented wave of demilitarisation and military budget cuts. Simultaneously, these same governments are increasingly opening their armies up to foreign nationals and outsourcing military operations to private companies. This book explores the impact of this hybridisation on the values, cohesion and effectiveness of the armed forces by comparing and contrasting the experiences of the French Foreign Legion, private military companies in Angola, and the merging of private contractors and American troops in Iraq.

Examining the employment of foreign citizens and private security companies as military forces and tools of foreign policy, and their subsequent impact on the national armed forces, the book investigates whether the difficulties of coordinating soldiers of various nationalities and allegiances within public-private joint military operations undermines the legitimacy of the state. Furthermore, the author questions whether this trend for outsourcing security can realistically provide a long-term and positive contribution to national security.

This book will be of much interest to students of private military companies, strategic studies, international security and IR in general.

Caroline Varin is a lecturer at Regent's University in London and has a PhD in International Relations from the London School of Economics.

Series: LSE International Studies
Series Editors: John Kent, Christopher Coker, Dominic Lieven and Karen Smith

The International Studies series is based on the LSE's oldest research centre and like the LSE itself was established to promote interdisciplinary studies. The Centre for International Studies facilitates research into many different aspects of the international community and produces interdisciplinary research into the international system as it experiences the forces of globalisation. As the capacity of domestic change to produce global consequences increases, so does the need to explore areas which cannot be confined within a single discipline or area of study. The series hopes to focus on the impact of cultural changes on foreign relations, the role of strategy and foreign policy and the impact of international law and human rights on global politics. It is intended to cover all aspects of foreign policy, including the historical and contemporary forces of empire and imperialism, the importance of domestic links to the international roles of states and non-state actors, particularly in Europe, and the relationship between development studies, international political economy and regional actors on a comparative basis, but is happy to include any aspect of the international with an inter-disciplinary aspect.

American Policy Toward Israel
The Power and Limits of Beliefs
Michael Thomas

The Warrior Ethos
Military Culture and the War on Terror
Christopher Coker

The New American Way of War
Military Culture and the Political Utility of Force
Benjamin Buley

Ethics and War in the 21st Century
Christopher Coker

Armed Groups and the Balance of Power
The International Relations of Terrorists, Warlords and Insurgents
Anthony Vinci

The Future of Biological Disarmament
Strengthening the Treaty Ban on Weapons
Nicholas A. Sims

The Character of War in the 21st Century
Paradoxes, Contradictions and Continuitues
Edited by Caroline Holmqvist-Jonsäter and Christopher Coker

America, the UN and Decolonisation
Cold War Conflict in the Congo
John Kent

Hamas and Suicide Terrorism
Multi-causal and Multi-level Approaches
Rashmi Singh

Mercenaries, Hybrid Armies and National Security
Private Soldiers and the State in the 21st Century
Caroline Varin

Mercenaries, Hybrid Armies and National Security

Private soldiers and the state in the 21st century

Caroline Varin

LONDON AND NEW YORK

First published 2015
by Routledge

2 Park Square, Milton Park, Abingdon, Oxfordshire OX14 4RN
52 Vanderbilt Avenue, New York, NY 10017

Routledge is an imprint of the Taylor & Francis Group, an informa business

First issued in paperback 2020

British Library Cataloguing-in-Publication Data
A catalogue record for this book is available from the British Library

Library of Congress Cataloging-in-Publication Data
Varin, Caroline.
Mercenaries, hybrid armies and national security : private soldiers and the
state in the 21st century / Caroline Varin.
    pages cm – (LSE international studies series)
    Includes bibliographical references and index.
    1. Mercenary troops–Case studies. 2. Private military companies–Case
studies. 3. Integrated operations (Military science)–Case studies.
    4. Armed forces–Case studies. 5. Military art and science–Case studies.
    6. State, The–Case studies. 7. Military policy–Case studies. I. Title.
    U240.V37 2014
    355.3'54–dc23   2014012178

ISBN: 978-1-138-77948-8 (hbk)
ISBN: 978-0-367-60042-6 (pbk)

Typeset in Times New Roman
by Wearset Ltd, Boldon, Tyne and Wear

Pour Nelly

This book has an impressive intellectual boldness and a striking theme: the hybridisation of the armed forces and its impact on the security of the state. One realises from a book like this that serious security studies must remain dependent not only on the political analysis of contemporary events but also on the insightful interpretation of history of just the kind that Varin supplies.

Professor Christopher Coker, *London School of Economics*

Contents

Acknowledgements

First and foremost, a heartfelt thank you to Christopher Coker for his invaluable support and friendship throughout this entire process.

To my father *"sans qui rien n'eut été"*.

To Chris Alden and the team at SAIIA, for their friendship, and for generously sharing office space, contacts and experience during my fieldwork in Johannesburg and Cape Town.

To those who read and commented on the early versions of this manuscript, Daniel, Chris, Claire-Marine, Calypso, Andrew and Ruben. I really appreciate your help and support!

And to the others, my friends and family: Mom, Aida, Michele, Miki and Marlen, Ian, Taty, Jo and Phillou, Steph and little Elysa – thank you for being my inspiration in London and around the world!

Abbreviations

BAPSC	British Association for Private Security Companies
CIA	Central Intelligence Agency
CEO	Chief Executive Officer
CPA	Coalition Provisional Authority
COIN	Counterinsurgency
DoD	Department of Defense
EUFOR	European Union Force
EO	Executive Outcomes
FAA	Forças Armadas Angolanas
REC	Foreign Cavalry Regiment
REI	Foreign Infantry Regiment
FMA	Foreign Military Assistance
REP	Foreign Parachute Regiment
FNLA	Frente Nacional de Libertacao de Angola
GAO	Government Accountability Office
IHL	International Humanitarian Law
IPOA	International Peace Operations Association
ISOA	International Stability Operations Association
ITAR	International Traffic in Arms Regulations
KBR	Kellog Brown & Root
MSC	Military Service Contracting
MSP	Military Stabilisation Program
MPLA	Movimento Popular de Libertacao de Angola-Partido do Trabalho
MAD	Mutually Assured Destruction
NCACC	National Conventional Arms Control Committee
FLN	National Liberation Front (Algeria)
NCO	Non-Commissioned Officer
NGOs	Non-Governmental Organizations
NATO	North Atlantic Treaty Organization
OAU	Organisation of African Unity
PCA	Partido Comunista Angolano
PMSC	Private Military and Security Company

PMC	Private Military Company
PSC	Private Security Company
PSP	Private Security Provider
ROCs	Reconstruction Operations Centres
RSLMF	Republic of Sierra Leone Military Forces
RUF	Revolutionary United Front, Sierra Leonean rebel group
SADF	South African Defence Force
SWAPO	South West Africa People's Organization
SAS	Special Air Service
UNITA	Uniao Nacional para a Independecia Total de Angola
UMHK	Union Minière du Haut-Katanga
UN	United Nations
UNIFIL	United Nations Interim Force in Lebanon
US	United States
USAID	United States Agency for International Development
SEALS	United States Navy's Sea, Air, and Land
WPPS	Worldwide Personal Protective Services

1   Introduction

The presence of foreign nationals and private contractors in contemporary armies has been increasing over the past two decades. Yet, the professional citizen-soldier has been the symbol of nationhood and state control, representing a society's cultural values and political ambitions for the past 200 years. The recent trend of states shifting towards a more open model, including foreign and private actors into the military body, has forced a hybridisation of the armed forces where soldiers are required to adapt, with more or less success, to this non-state agent. The composition of the national armed forces in the twenty-first century is going to continue to include more foreigners, contractors and soldiers, working side by side to further the interests of the state.

While this phenomenon has already gained much academic and political interest, it is worth adding to the literature with a new comparative and historic study of the hybridisation of the armed forces and its impact on national security. In particular, including new actors into the army is challenging the status quo of the military institution, an organisation that is traditionally rigid to change, relatively homogeneous, and with a set modus operandi. This book therefore asks three main questions:

1   What are the differences, if any, in the ability of foreigners, private contractors and citizen-soldiers to guarantee the internal security of the state and successfully carry out its foreign policy?
2   What happens to an army of citizen-soldiers when it is forced to work side by side with private, sometimes foreign companies?
3   In the long run, can the hybridisation of the armed forces actually improve, or does it impede, the security of the state, and what efforts are being made to facilitate this transition?

This is a particularly important topic in view of the increasing number of flailing states where weak leaders are outsourcing their security needs to foreign and private actors in order to prop up their own regimes. In the last twenty years, the governments of Papua New Guinea, Angola, Liberia and Sierra Leone have all turned to foreign private military companies to train their armies and supplement their troops against rebel insurgencies. In 2004 and again in 2011, former

President Laurent Gbagbo called upon Liberian mercenaries to preserve his rule in Côte d'Ivoire. Similarly, former Libyan leader Colonel Muammar Gaddafi recruited between 5,000 and 10,000 principally Touareg mercenaries from Mali, Niger, Chad and Algeria to quell local uprisings that challenged his right to rule, to no avail. Western states, following unprecedented budget cuts to the armed forces, are also outsourcing their military capabilities to foreigners and private contractors in the Middle East and in resource-rich African countries where they have significant investments. Even China has begun to use private security companies "staffed with 'retired' members of China's security forces" (*FT* 2/2/2012) to provide security to its commercial operations in conflict zones, notably in the Sudan and other African countries.

Considering this trend and its lack (but not total absence) of historic precedent, it is important to regularly reassess the viability to the model, taking into account similar cases where armies have had to adapt, for better or for worse, to new roles, security threats and combatants. This work, therefore, specifically addresses the question of how the hybridisation of the armed forces has affected national security in an increasingly integrated and competitive political environment.

Defining combatants

Defining the differences between types of combatants is challenging, as these actors vary both from state to state and from each other. There is no consensus as to a definition of 'mercenary' or 'soldier', although both combatants have been used extensively throughout history, from the Greeks who were notorious as the best mercenaries of their epoch[1] to the Napoleonic armies of supposedly citizen-soldiers. It is possible to differentiate at least three types of combatants for our purposes: 'citizen' soldiers, pooled from the nation's citizenry and with a vested interest in the security of their own state; foreign soldiers, such as the Gurkha Regiments and the French Foreign Legion who are incorporated into the armed forces but often remain marginal actors under the strict supervision of national officers; and private contractors, sometimes amalgamated with 'mercenaries', who are external to the military institution of the state but can be contracted by anyone to participate in a conflict.

Each of these groups differs in motivation, goals and organisational structure. They have also been more or less welcome as providers of security in the development of the nation-state. For example, there are both national and international laws that forbid the recruitment of foreigners into the armed forces, and likewise, make it illegal for nationals to enrol in foreign armies. In the Middle Ages, however, professional mercenaries were perceived as more trustworthy combatants than serfs. As norms change, so do the combatants in warfare.

Using the term 'mercenary' to designate non-state combatants has also become politically charged and heavily contested, not least by academics. Peter Singer distinguishes mercenaries by their foreignness to the conflict, their independence from the national force and the limitations of the contractual ties, their short-term economic motivations, the method of recruitment and organisation and the nature of their services. Janice Thomson defines mercenarism as "the practices of enlisting in and recruiting for a foreign army" (Thomson 1994: 27), and entirely leaves out the motive argument on the basis that "individual motives are impossible to determine". Sarah Percy, on the other hand, focuses exclusively on the financial motives of mercenaries and their lack of personal interest in the conflict: "mercenaries are morally problematic because they cannot provide a plausible justification for killing; they cannot point to a cause in the service of which they fight, aside from financial gain" (Percy 2007: 54). The Italian philosopher Niccolò Machiavelli corroborates this point by defining mercenaries by their lack of motivation in combat: "they have no tie or motive to keep them in the field beyond their paltry pay, in return for which it would be too much to expect them to give their lives".

Many of these points distinguish mercenaries from soldiers in terms of their relationship vis-à-vis the nation, i.e. their nationality. It is assumed that only citizens have an intrinsic interest in the security of their land, while foreigners have other motivations which can easily fluctuate – hence the classification of these combatants into 'soldiers' with an assumed legal and emotional attachment to their country, and 'non-state warriors' who are *a priori* external to the state. The perception that combatants 'must' be nationals is a social norm born out of the evolving function of the state and not necessarily a reflection of the actual efficiency and trustworthiness of each combatant.

Violence and social norms

A fundamental element of war, the choice of combatants is susceptible to normative preferences that reflect a society's acceptance of who can yield violence and how. The European tradition of civic militarism and the definition of the modern state in terms of its monopoly over the legitimate use of violence have influenced contemporary social norms determining the role of mercenaries and soldiers in foreign policy. Alasdair MacIntyre explains that war takes place "within the context of norms which a community shares" and therefore the culture of war changes geographically and historically as norms can vary from place to place. For the past 200 years, on the European continent, citizens have been the preferred combatant, representing the cultural values and choices of the state and society.

Social norms define everything: who can exert violence, who can kill, who can be a soldier, and so forth. Norms are "a set of rules that stipulate the ways states (and people) should cooperate and compete with each other. They prescribe acceptable terms of (state) behaviour and proscribe unacceptable kinds of behaviour" (Mearsheimer 1994). They are important because they represent a society's ethos and values and subsequently can influence their state's choices and behaviour. In a comparative study of combatants, the choice of using non-state warriors – also selectively referred to as 'mercenaries' in this study – or soldiers in wars is contingent on the social norms that envelop the level of acceptability for each actor.

The actual influence of norms, however, has been keenly debated in international relations. John Mearsheimer argues that norms are "a reflection of the distribution of power in the world. They are based on the self-interested calculation of great powers and they have no independent effect on state behaviour". They are also created by the state to guide the value system of their citizens and nurture their ideological and political support. This presupposes that the interest of the state actually primes over value systems that in turn can be moulded and changed according to the whims of the state. Norms can be influenced exogenously, however. Significant changes in military technology and in the scale of war have forced armies to adapt in order to survive, regardless of the prescribed norms and preferences of the state and society. Constructivists, on the other hand, argue that norms are an intrinsic part of state identity and therefore determine state behaviour. This explains why states sometimes pursue policies that appear counter-productive to their immediate interests and aspirations. Sarah Percy claims that the constructivist approach is "best suited to explaining the norm against mercenary use (...) given that state interests of the desirability of deploying private force have changed enormously, in ways that cannot always be accounted for by material factors" (Percy 2007: 18).

Norms are perpetually changing to accommodate new innovations and paradigms that "unsettle existing structures of knowledge about the past and its relation to the present" (Coker 2010: 142). Philosopher Thomas Kuhn suggests that paradigm shifts "occur not in the minds of individual innovators, but in particular conjunctures of social and intellectual circumstances which challenge existing structures of knowledge and open up space for new ideas". War, adds Coker, is especially susceptible to these paradigm shifts because it is itself "the invention of culture", reflecting a society's values and identity and its adaptability to changes in its immediate environment. Because "war is anchored to what we imagine or would like it to be, it is in that sense profoundly normative" (ibid.: 144).

The current preference for soldiers and the concomitant condemnation of mercenaries, however, have not been a constant throughout history. Until at least the nineteenth century, states conducted their foreign policies through non-state subsidiaries including mercenaries, mercantile companies and privateers. The normative hostility towards non-state combatants stems from two social norms: one which argues that "mercenaries were antipathetic to the norms of modern community" (Coker 2010: 149) due to the instrumentality of war for social conditioning and the foreign and therefore apolitical nature of mercenaries who are not part of the established community. The second claims that the state must maintain a monopoly over the legitimate use of violence and be capable of holding all military actors accountable (this is further developed in Chapter 7). Mercenaries are foreign, freelance, and therefore very difficult to control, making them undesirable in the eyes of modern society. These norms, it will be shown, were deliberately designed and adapted by the state in its pursuit and consolidation of power and legitimacy.

Research design

This project was born out of a desire to assess the viability of alternative security options in war-torn African countries. Taking security as a necessary premise for development and factoring in the reluctance of certain regimes to reinforce their national security apparatus, can private military providers offer a tangible solution and improve the security in a given state? In order to assess this possibility, it was first necessary to tackle the normative argument that foreign combatants are, if not always a direct security threat, an undesirable agent in contemporary warfare. A thorough review of contemporary and historic case studies further enabled the study to evaluate the risks of hybridisation, i.e. integrating soldiers with non-traditional warriors in one same force. By investigating trends of hybridisation in similar but not equal situations, it was possible to evaluate the viability of this model and make suggestions for states that choose to develop their military apparatus in this direction.

Chapters 2 and 3 compare and contrast the historical role of soldiers and non-state warriors and evaluate their respective abilities as combatants – the first question that we seek to address in this book. Chapter 2 explores the historical use and value of mercenaries in warfare prior to 'Classical Modernity' (Berman 1982). Mercenaries pre-date soldiers, and were used by warlords and sovereigns to pursue their territorial ambitions. The fashion of using mercenaries was challenged by the French Revolution in 1789 and the ensuing rise of nationalism that advocated in favour of citizen-soldiers as the more dependable, desirable and controllable combatant. Motivations, morals and military virtues are explored in this chapter through a collection of interviews, philosophical arguments and battle assessments that reveal that mercenaries are not solely motivated to go to war by the lure of profit, nor do they necessarily make worse combatants than soldiers. Nonetheless, their continued lack of accountability and restraint distinguishes them legally, morally and psychologically from state-warriors, fostering tension and resentment within the ranks when they are forced to work together.

Chapter 3 focuses on the soldier, and presents his rise to prominence as the state's chosen combatant in the nineteenth century, following France's Revolutionary Wars and the military reforms of the Prussian Army. The military institution came to represent the values, strength and aspirations of the modern state while civic-militarism, duty and sacrifice were deliberately instilled in the population through massive state propaganda campaigns and through education. Patriotism and nationalism were presented as necessary attributes defining the moral and military value of the soldier who in turn received honour and recognition from his society. Obedience was insured through draconian training inspiring blind obedience, and a system of military courts that punished deviants and discouraged defiance. These measures guaranteed the accountability and performance of the soldier which may not have been ensured otherwise. Through a survey of historic and more recent academic and literary accounts, this chapter contrasts the efficacy in combat of mercenaries, who elect to join a conflict and remain outside of state structures, with that of soldiers, who are bound to and by the state.

The hybridisation of the armed forces is a relatively recent phenomena, with only a few cases to choose from in order to assess its impact on civil–military relations (i.e. the relationship between soldiers and non-state actors). Consequently, it was necessary to delve into like cases where national soldiers have been required to collaborate with foreigners integrated into their ranks. The French Foreign Legion was chosen as the obvious case study, with a rich literary history available in several languages that reflects the difficulties and advantages of merging French soldiers with foreign combatants in the pursuit of French foreign policy. Although legionnaires, like the Gurkha, are not uniformly regarded as mercenaries, the individuals participating in combat are foreigners with no obvious emotional attachment to the country for which they are fighting, and are for our intents and purposes categorised as pseudo-mercenaries. Cognisant of this difference, national soldiers, and indeed society as a whole, have regarded legionnaires as marginal and sometimes suspicious agents and have forcibly limited their enrolment in the nation's armed forces.

Chapter 4 therefore looks at the French Legionnaire as a quintessential example of the merging of military values and mercenary attributes. It traces the military experiment of creating a regiment of foreigners from its difficult start as a funnel for immigrants and undesirables, to its successes and failures in the colonial and post-colonial military campaigns. Despite initial obstacles and national resentment, the Foreign Legion progressively became the striking arm of the French colonial project, and the pride of the French Army with its own international reputation. The chapter also explores the identity and motivations of the men who join as legionnaires and the training and disciplinary system of the Legion, with the aim of understanding the process by which these mercenaries were transformed into respectable and effective 'soldiers'. As a mercenary unit, the Legion suffers from above average desertion and suicide rates. Its members, all volunteers, have come from extreme backgrounds, ranging from criminals to intellectuals. Despite their differences, legionnaires exhibit unusual group cohesion and loyalty to the Foreign Legion, which has played out on the battlefield to create a fierce and dependable force that carries out its orders, facilitating French foreign policy.

The case study of the French Foreign Legion is particularly useful as it dispels the assumption that foreign warriors cannot be reliable combatants or function effectively with the national armed forces. It is also useful as a model of how 'mercenaries' can be controlled when incorporated into a state structure. The success of the Foreign Legion lies not only in its mythical reputation and stringent training and disciplinary programme, but also in its integration into the French *Armée de terre*. It remains, however, under direct French legal and political control, ensuring the accountability of the legionnaires.

African states have perhaps been the theatre with the most exposure to mercenaries and are arguably most in need of an external force-multiplier to improve security and reduce the prevalence of conflict in their territory. Africa, however, has had mixed results in its dealings with foreign and private combatants, and is therefore the subject of Chapter 5. The chapter begins by contextualising the variables that led to a breakdown of security in several states. In particular, it evaluates the impact of tribalism, colonialism and post-colonialism on the identity of the armed forces and on their relations with the civilian government and the population. The security environment of post-colonial Africa, with its weak institutions and unreliable armies, created a vicious circle of violence, fuelling conflict and hindering socio-economic and political development. Mercenaries were subsequently hired by secessionist states, ousted leaders and independent entrepreneurs to illegally challenge the legitimacy of the government. These private armies of mercenaries gave their employer a military advantage that went beyond their capabilities, and could potentially turn the tides of war, threaten the establishment, and challenge the status quo, but perhaps also improve the security situation. Consequently, African leaders have been especially pro-active both in passing legislation condemning mercenaries, and in hiring these non-state actors when 'necessary' to the survival of their regime. The use of these non-state combatants, however, has been very contentious at the level of the national armed forces, because of their foreignness and lack of accountability and interest in the state.

The case study of Angola is salient as a well-documented model of a successful, albeit short-term military collaboration with a private company in the context of a civil war, and can serve as a model for other African states seeking to outsource security or integrate non-state warriors into their military apparatus. Angola was also selected in this comparative study because it is arguably the best documented example of civil–military cooperation, after many political actors and former mercenaries wrote about their experiences and perceptions in the conflict. These same actors related their thoughts and opinions on the hybridisation of the armed forces in a series of interviews for this book. The interviewees showed that, while privatisation might appear as a short-term solution to civil conflict, it is fraught with difficulties and can ultimately be damaging in the long term to the identity, loyalty and hierarchical structure of the national armed forces and the security of the state. This is particularly the case when governments opt to pour their resources into short-term solutions in the form of a private military company rather than investing in the military infrastructure of the state.

Finally, no Western country has been more aggressively outsourcing security to private military companies than the United States, especially in the wake of the Iraq War. The emergence of private and foreign combatants as a key component of US military strategy has created new opportunities in US foreign policy, and exacerbated old tensions between non-state actors and the professional armed forces. As Europe pursues the American model of hybridisation, it is important that it is made aware of the dynamics involved in this process.

Chapter 6 explores the rise of the private military sector within the context of privatisation and outsourcing that already characterised US domestic policies. Using Iraq as a case study for America's experiments in military public–private partnerships, the chapter outlines the advantages and difficulties of hybridising the armed forces and focuses on civil–military relations, issues of command and control, overlapping identities and blurred hierarchies. The massive outsourcing of logistics, tech support and even military tasks in the last decade has changed the face of security in Iraq, with contractors outnumbering regular troops. Military operations are increasingly supported or even replaced by private military and security firms, leading to a difficult cohabitation between the two actors that can adversely affect the war effort. The lack of accountability remains an obstacle, however, impeding the integration of contractors as a regular feature of contemporary warfare. Contractors remain beyond the control of the state, and as such are viewed with justified suspicion by the local population and US military personnel, despite often being of the same nationality as the soldiers next to whom they are fighting, and regardless of their perceived usefulness to the ambitions of the government.

Chapter 7 focuses on the evolving role of the state and its relationship vis-à-vis its chosen combatant, using the evidence found in the three afore-mentioned case studies. This is necessary to evaluate how the hybridisation of the armed forces in different situations can affect the traditional dynamics connecting the tripartite state, i.e. the government, the army and the population.

The chapter begins with an overview of the changing role of the state and its reliance upon armed combatants to establish legitimacy and hold on to power. It analyses the use and manipulation of norms by the state, and shows how these norms can be adapted to suit the purposes of the ruler, particularly regarding the waging of wars. As the state's need for combatants exceeds its domestic supply and capabilities, it has historically turned to foreigners, mercenaries and corporations to meet its demand, and has easily justified this shift to its citizens. By outsourcing its military needs to non-state actors, however, the state has reached a new zenith of power, where it can shift moral and legal responsibility away from itself and evade the system of democratic control that the national army claims to safeguard. Taking into account the lack of accountability that has until recently characterised non-state actors, the chapter concludes that mercenaries who are hired by a state uphold the power and centrality of said state, despite accusations to the contrary. On the other hand, they also threaten established norms of democratic accountability by giving the state the ability to act beyond the will of the people. Furthermore, they undermine the morale and, ultimately, the strength and performance of the national armed forces.

Finally, the concluding chapter revisits the empirical evidence within the context of the conceptual framework. It highlights problems that have recurred in the case studies, and suggests policy options to inform politicians and military decision-makers on whether and how to hybridise the national armed forces in the pursuit of an efficient and effective security policy.

Note

1 Alexander the Great made a point of killing every Greek mercenary whom he came across as he considered them to be the biggest threat to his military campaigns.

2 Non-state warriors

The mercenary's role in war is fraught with controversy, illegitimacy and condemnation. Yet mercenaries have been hired time and again to carry out a military objective for their respective employers. Not unlike the soldier, the mercenary is a romantic, lone figure who has caught the imagination of the world and inspired countless tales of prowess and intrepidity. The mercenary is the ultimate adventurer, rootless and independent, whose ability and willingness to kill for others comes at a price. But he is a horribly mutated type of combatant: his isolation, fearlessness, apparent lack of morals and constant presence in the wars of the world have made him into a misunderstood, magnetic figure that is both the embodiment and antithesis of humanity. Mercenaries are *not* soldiers, and this distinction is probably the root of the difficulties that define the relationship between these two combatants on the battlefield.

This chapter analyses the function of mercenaries in warfare. It briefly develops historic case studies where mercenaries have played a decisive role in the projection of their employer's domestic and foreign policies: from the development of Greek mercenaries into the most notorious force of their era to the mercenary revolt that led to the fall of the city-state of Carthage, the use of Swiss Pikemen in the armies of Medieval Europe and the success of the White Company in Italy. The narrative then turns to the rise of nationalism in the eighteenth century and its impact on military and political consciousness which changed the ways in which mercenaries have been perceived and employed. Modern objections to mercenaries and mercenarism are subsequently explored and critically evaluated against a thorough psychological profiling of these non-state combatants. This chapter therefore provides a historical and political snapshot of mercenaries and serves as a reference for analysing their similarities and differences with the more conventional soldiers.

The historical value of mercenaries

Despite the normative assumption that mercenaries are intrinsically corrupt and should have no role in war, these rogue combatants actually pre-date soldiers in their historic place on the battlefield, hence their reputation as the second oldest professionals in the world after prostitutes. Historian Anthony Mockler makes the case that mercenaries only became disreputable with the emergence of the nation-state in Western Europe and the introduction of universal conscription in the nineteenth century. Azar Gat refers to this paradox as the "non-state armed bands in a state environment" (Gat 2006: 265), where mercenaries challenged the ambitions and goals of the new political system.

Even so, until the nineteenth century, states mainly conducted their foreign policies through non-state subsidiaries: mercenaries, mercantile companies and privateers. No history of warfare would be complete without an expansive description of the role played by mercenaries. Indeed, hiring mercenaries was the norm, and not the exception, as powerful states preferred to outsource their security requirements to foreigners rather than send their active citizens to war. Most sovereigns required their citizens to participate economically in the development of the state and contribute to the tax base. They were also averse to arming their subjects in case these chose to revolt against the powers that be. Consequently, the peasantry were burdened with labour-intensive work, merchants maintained a healthy repugnance for war, and although the elite classes abided by a certain *noblesse oblige*, they were averse to exposing their sons to real danger. Sovereigns therefore turned to mercenaries as a practical solution to enforce their rule through the use of organised armies and to project their power by means of violence. Even where rulers maintained forces of citizen-soldiers, these were often too small to both secure the state and travel abroad to invade and control new territories. Ambitious leaders subsequently began to recruit men from neighbouring states to complete their armies.

Early mercenaries were rarely organised. They entered into the employ of another state as individuals, and were subsequently grouped with other individual mercenaries to form a battalion. Entire armies functioned on this model. Mercenaries sometimes had a contract, but were generally considered free men and could break the contract at any time. They had to supply their weapons and armour themselves and were often responsible for their own subsistence, which further pushed them towards pillage, theft, desertion and rebellion.

Ancient mercenaries

The oldest known case of foreign military assistance can be dated 3,000 years ago in Ancient Egypt. The Egyptian population was employed in agriculture and building monuments for the glory of its rulers, and hence was neither trained nor available to fight wars. The Egyptian Empire had to resort to recruiting warriors from the regions bordering its territory to fight its wars. Therefore, to expand their empire, the pharaohs Sesostris III (1878 BC–1843 BC) and Ramses II (1279 BC–1213 BC) hired Nubian, Palestinian and Syrian mercenaries. Greek and Asian mercenaries served in Psammetique I's army and expelled the Ethiopians and Assyrians from Egypt. In the subsequent empires, Egypt incorporated foreign soldiers from its conquered lands into the army: Pharaoh Apries is believed to have hired an army of 30,000 mercenaries, largely made up of Greeks, who were the most sought after warriors in Asia Minor. Indeed, traces of Greek hoplites dating from the sixth century BC can be found on the Colosses of Abou Simbel, a testimony to the large presence of foreigners in the 'Egyptian' armies.

The notoriety of the Greek mercenaries is engraved in the famous Expedition of the Ten Thousand. In 401 BC, the Persian prince Cyrus the Younger enrolled more than 13,000 Greek mercenaries to fight against Artaxerxes, the King of Persia. The mercenaries were hired to travel from Sardinia to Babylon merely to fulfil the personal ambitions of the prince, who was eventually killed in Counaxa.[1] Under the leadership of the soldier Xenophon, and against all odds, the mercenaries retreated back to Greek territory, pillaging the countryside on the way. Xenophon recorded the events in his book *Anabasis*, which inspired Alexander the Great in his expedition into Persia. The Macedonian king reinforced his army with mercenaries who were left unemployed by the Greek wars and marched over 44,000 of them into Asia Minor in 334 BC. According to historian Victor David Hanson, he also was infamous for massacring any Greek mercenary he found fighting for the 'barbarians' on the other side:

> Alexander may have exterminated between 15,000 and 18,000 Greeks after the battle (of Granicus) was won. (...) Alexander would have to kill like no other Westerner before him to achieve his political ends, and he would be forced to eliminate thousands of Greeks, who for either greed or principle were willing to fight him in service of the Persian king.
>
> (Hanson 2001: 82)

Through their campaigns and military successes, Greek mercenaries acquired a reputation that transcends history: "the murderous Hellenic-inspired armies – the Ten Thousand, the Macedonians under Alexander the Great and the mercenaries of Pyrrhus – possessed of superior technology and shock tactics, would run wild from Southern Italy to the Indus River" (Hanson 2001: 82) and would determine the fate of battles for centuries.

Carthage and the mercenary wars

On the other hand, the case study and eventual defeat of Carthage by the Romans in the third century BC offers the first signs of what would become a tradition of civic-militarism – "the notion that those who vote must also fight to protect the commonwealth, which in the exchange had granted them rights" (ibid.) – and early signs of a declining reliance on mercenaries. Carthage was a rich city-state based on the coast of what is now known as Tunisia. It was also the most important rival to the expansionist ambitions of the Roman Republic, and the two powers confronted each other in three successive wars that became known as the Punic Wars. At the time, the Roman Legion was a standing army made up of Roman citizens, whereas the Carthaginian city-state was a heterogeneous amalgamation of nationalities and the army was complemented by at least 40,000 mercenaries.[2] The rivalry between the two powers reflects the political choices and military effectiveness of hiring mercenaries (on the Carthaginian side) versus enlisting citizen-soldiers (on the Roman side).

The First Punic War lasted 20 years, from 264 BC to 241 BC, at the end of which Rome, with her citizen-army, was victorious and imposed heavy reparations on Carthage who was unable to pay its mercenaries returning from battle. Predictably, the mercenaries assembled together and seized Tunis, demanding payment for their services and refusing to negotiate with the Carthaginian representatives. The mercenaries were further supported by the Libyans, led by Matho, who were also rebelling against the oppressive rule of Carthage. Although the revolt was eventually crushed by the Carthaginian General Hamilcar Barca,[3] distracted by its internal problems, Carthage was unable to prevent Roman expansionism and thus lost its territories of Sicily, Sardinia and Corsica, which enabled the finalisation of Roman monopoly in the Mediterranean (her *Mare Nostrum*). The mercenary wars had important ramifications for the balance of power in the region. Carthage's eventual defeat in the Punic Wars, according to Machiavelli, can largely be blamed on the mercenary revolt, which took up resources and weakened the state and its military branch: "Carthaginians at the close of their first war with Rome were well-nigh ruined by their hired troops, although these were commanded by Carthaginian citizens" (Machiavelli 1992: 32).

Victor Davis Hanson stresses this point by stating that the civic-militarism of Rome facilitated a continuity of military campaigns and victories that goes beyond its army's defeat at the Battle of Cannae in 216 BC. Indeed, the lesson for Hanson is that "students of war must never be content to learn how men fight a battle, but must always ask why soldiers fight as they do, and what ultimately the battle is for" (Hanson 2001: 128). By hiring mercenaries, Carthage might have increased her military power temporarily, but at huge risk and high cost as these mercenaries were unreliable and ended up rebelling against the hiring state, causing more problems. Rome, on the other hand, was able to recover from its losses through a campaign of mass mobilisation of citizens in which every able man was drafted into the militia. The mobilisation of Roman citizens and the Republican tradition of citizen-warriors eventually led to Carthage's defeat in 146 BC, and Rome ruled over the civilised world for a further 500 years. Ultimately, the end of the Roman Empire is attributed to the loss of its tradition of civic-militarism and its growing military dependence on barbarian troops which made up the bulk of Roman forces by AD 476, the date that historians mark as the end of antiquity. The fall of the Roman Empire provoked a return to mercenary armies in Western Europe, but not without leaving a legacy of a warrior identity and a tradition of civic-militarism that would re-emerge in the eighteenth and nineteenth century.

The Swiss Pikemen

The wars of the Middle Ages are characterised by the overlapping identities and allegiances that marked the feudal system and left little distinction between a paid soldier and a mercenary. The concept of foreignness only began to emerge at the end of the thirteenth century in "independent and increasingly centrally administered states where distinctions between local, 'national', 'own' troops and 'foreign' troops became gradually apparent" (Keen 1999: 2010). The Swiss cantons were exactly that.

In the fourteenth century, Switzerland was a loose confederacy of poor cantons. Its inhabitants were accustomed to a harsh climate and difficult terrain that offered few natural resources and even less economic opportunities for their sons. Young men unable to find a trade were oriented towards the military arts, which only contributed to exacerbating the tensions between the cantons which were constantly at war with one another. United in a common goal to eject the Austrians from their country, the cantons crushed the mighty Habsburg armies at the famous battles of Morgarten (1315), Sempach (1386) and Näfels (1388). These military victories against the vastly superior Austro-Hungarian Army earned the Swiss Pikemen, so called because they carried pikes, a reputation as the most fearless soldiers in Europe.

Thereafter, Swiss mercenaries became a pillar of European wars, present in every battlefield and fighting on all sides. The humiliation of the Habsburgs delighted their enemies and created a demand for Swiss soldiers to serve in the armies of European kings: in 1480, the Cent Suisses were commissioned by the French King Louis XI to serve as his personal guard in the Louvre Palace. By 1567, the cantons guaranteed a permanent Swiss Guard Regiment to the French sovereigns and King Francis the First hired a further 120,000 Swiss mercenaries to supplement his military campaigns. Between the seventeenth and eighteenth centuries, the model of the Cent Suisses was reproduced in most of Europe's royal families, from the Kingdom of Sardinia to the Dutch Republic. The Swiss Guard that today still protects the Pope in Rome has been continuously employed through a pact with the cantons that dates back to 1471. Thus, between the fourteenth and the eighteenth century, an estimated one million Swiss mercenaries served abroad, particularly in France and in Italy, where the Vatican's Swiss Guard remains the last vestige of the Swiss Foreign Service.

The ferocity and courage of the Swiss mercenaries even earned them the praise of Machiavelli, who commended their sense of freedom and equality and viewed the Swiss militia system as a model for others. With a high demand for Swiss mercenaries, a new market was opened to the cantons who began hiring out their excess men in exchange for a tidy profit: "the soldiers were paid four florins a day, but only for so long as their service lasted, whereas the contracts between prince and canton were normally for a minimum period of at least one campaigning season" (Mockler 1970: 86). This enabled the canton of Valais, for example, to finance its expenditures without raising any taxes on its citizens. Modern Switzerland can credit its system of organised mercenaries for propelling the country out of poverty and into the military history annals of Europe.

Although mercenaries were sent out to fight for other countries, they remained under Swiss jurisdiction. As early as 1394, all cantons agreed to the Swiss military code in which they were given the duty to supervise the military training of their men, thus starting a long tradition of military service. The men who went to fight for foreign lords had to make an official military oath to the Confederation, and they represented their country for the entire duration of the campaign. The Swiss mercenaries largely acquired a reputation for being loyal and reliable. They often fought to the death: in particular, in 1527, the entire Swiss Guard in the Vatican was killed defending Pope Clement VII as Spanish mercenaries sacked Rome. In its most famous episode in history, 600 of the 900 Swiss Guards defending the French king at the Tuileries Palace on 10 August 1792 were massacred by the Parisian crowd.

Although Swiss mercenaries were naturally motivated to 'enlist' by the lack of economic options in their rural cantons, they also famously nurtured a love for combat and warfighting. Indeed, the Swiss were constantly at war, if not with foreign powers, then among the cantons themselves.[4] David Courtwright argues that "whenever young, single men congregated for long period under other than stringent discipline, violence ensued" (Courtwright in Tilly 2003: 1). A surplus of armed and unemployed men has always been considered a threat to the security of the state. As a result, communities repeatedly 'encouraged' their excess population to migrate to new opportunities so as to control the level of indiscipline and potential for violence inside their territory. This model was evident in the case of Xenophon's 10,000 mercenaries who, upon their return from fighting Cyrus' wars, found a cold reception as the Greek cities were reluctant to accept these soldiers of (mis)fortune and integrate them back into society. To be rid of the presence of these offensive men, the various Greek cities sent them to fight with the Spartan troops against the Satraps. This strategy was emulated after the Hundred Years War by the French who sent the roving bands of mercenaries across the border to fight in Spain. In a further bid to bring an end to the lawlessness, Charles VII of France made a first attempt in 1445 to integrate the mercenaries into the Royal Army. Finally, Switzerland perfected the administration of its excess young men and even managed to turn a profit from their military potential. The successful business of mercenarism developed by the cantons was soon imitated by other countries. Maximilian I, emperor of the Holy Roman Empire, became the biggest competitor to the Swiss enterprise by hiring out his Landsknecht whose heirs, the Dragons, were later contracted out to Great Britain to fight in the American Revolution. Eventually, mercenaries came together in their own companies to provide military services to foreign princes.

The case of the White Company

The successful business model of hiring out mercenaries that the Swiss cantons developed was first reproduced in a privatised form by Roger de Flor, who created the Great Company in 1302. The first company's modus operandi served as a blueprint for what eventually became the White Company, the precursor to modern private military companies.

The commercialisation of warfare in the fourteenth century was the result of the military organisational shortfalls of the feudal system, financial opportunities (especially in the Italian city-states), a vast supply of available soldiers, and a growing demand for non-state combatants. In particular, Italy offered "a vast marketplace where men in search of adventure and reward could strike profitable bargains with an employer" (Cooper 2008: 61). The Italian city-states preferred to hire foreign combatants to defend their territory and further their political interests because

> the fractious nature of political and social life caused local officials to think carefully before hiring their own for military service. A successful native captain might seek political gain at home; an unsuccessful captain might provoke scandal and internal quarrels between his supporters and opponents
> (Caferro 2006: 71)

Mercenaries therefore were considered to be more loyal than citizen-soldiers. William Caferro explains that hiring foreigners was a safe bet because, in theory, they had no interests in the politics of the city and could be easily fired. Hiring private companies instead of bands of individual mercenaries provided the contracting state with a professional service led by a reputable captain who could organise a military campaign and satisfy the security needs of the state in a structured framework, thus offering a certain amount of accountability and control. It also enabled the state leaders to dictate policy without fear of military retribution or need for public support. This particular environment fermented the corporate structure of mercenaries, which was to mature and evolve along the centuries adapting to the changes in the international system and the demands of war.

The White Company is the best documented example of early private military companies. Led by the famous Englishman John Hawkwood, the company emerged from the vestiges of the Great Company that crossed into Italy in 1361 during a lull in the Hundred Years War. The foreign mercenary bands amalgamated into companies that "were corporate in structure. The captain stood at the head of his brigades in a manner similar to the way a modern CEO stands at the head of his firm" (ibid.). They referred to themselves as 'societies', employed accountants, treasurers, secretaries, constables and lawyers, and were under the jurisdiction of their captain. When under contract, they also had to respect the conditions and the laws of the contracting party: "the Condotte provided that the leaders of the companies are responsible for the crimes committed by the soldiers in camp, while the Commune shall judge those committed in the city or to the damage of the republic" (Cooper 2008: 61).

In his first campaign with Pisa in 1363, Hawkwood had 3,500 horsemen and 2,000 infantry under his command. The companies signed contracts with various actors, mostly states which hired them to defend their territory and to attack their enemies, but private citizens also used the companies as escorts through the insecure routes of Northern Italy. At a time when mercenaries were perceived with suspicion and accused of pillage and intimidation, Hawkwood gained a reputation as an honest entrepreneur, characterised by his "steadfastness in performing his obligations to his employers" (Caferro 2006: 16). This view is supported by

Machiavelli in *The Prince*, although the Florentine chancellor insisted that Hawkwood's faith was never tested as he generally failed in his military campaigns. Hawkwood was generally considered as "a loyal soldier during the periods in which he was directly employed by a state, pope, or prince" (Mockler 1970: 63). The captain's reputation determined the success of his company. Hawkwood was a popular captain, having been elected to the position by his colleagues and inspiring the loyalty of his troops: he never suffered a mutiny at a time when indiscipline was a problem. Likewise, Italian city-states competed for the services of the White Company, although this may also have been motivated by their desire not to be the victims of a contract between the company and an enemy state, and to remain outside of the path of pillage, plunder and bribery that all companies, including the White Company, resorted to when not restrained by contractual conditions.

When the White Company was contracted by the city-state of Pisa in 1363, Florence and Pisa had already been antagonists for several hundred years due to Florence's ambitions to gain access to the sea through Pisa's port. Pisa in the thirteenth and fourteenth century was a rich city with successful commercial enterprises that enabled it to make up for its military weaknesses by securing a private army. The White Company served Pisa and, despite losing a few battles (including one against Florence in 1364), was arguably a stabilising influence in the foreign policies of the city-state: exempted from military service, the inhabitants of Pisa could direct their time and resources towards developing the economic and political infrastructures of the city. Their defences were ensured by the White Company, which deterred other cities from attacking and encouraged some city-states, particularly Florence and Milan, to make alliances with Pisa.

The White Company not only provided relative peace and encouraged prosperity in Pisa, it also enabled its employers to act out a foreign policy independently of foreign interference. During his unusually long career, Hawkwood and his White Company signed contracts with Pisa, then with Pisa's enemy Florence where he commanded the city's army and served as a police force for the Florentines during the last 14 years of his life, but also with the Milanese Duke Bernabo Visconti – whose daughter he eventually married, and with the Pope, among others. The services of the company, once hired by a state, could be subcontracted by the state to its allies in times of peace. Hawkwood also developed a network of spies throughout Italy which, along with his personal knowledge of the terrain and politics, enabled him to gain an advantage over competing companies and sell his information services to foreign and local dignitaries. The White Company was different from previous manifestations of mercenaries: it was a very modern concept that set a precedent by its corporate nature at a time when the European continent was witnessing the first seeds of capitalism.

As these case studies have shown, mercenaries were a normal and recurrent figure in the military landscape of Europe. Until the White Company, mercenary companies were a pre-commercial and loose organisation. Hawkwood's enterprise was a commercial outfit that represented the changing trends in European society. This began to change, however, with the rise of nationalism and national consciousness that spread across the continent, inspired by the French Revolution and the Napoleonic Campaigns.

The French Revolution and the rise of nationalism

Mercenaries were a fixture of sovereign power in the seventeenth and eighteenth centuries, and nowhere more so than in France. Mercenaries offered a practical solution to the limitations of feudalism which required men to remain under arms for only forty days. As kings turned to foreigners to arm their military campaigns, the armies of Europe became increasingly heterogeneous as multinational bands of mercenaries developed a permanent presence in the country that employed them. When Charles VII of France established the fifteen companies of a standing army, two of them were made up exclusively of Scots. By 1789, the French Army also included 11 Swiss regiments. These personal guards came to represent the foundations, ambitions and, subsequently, the tyranny of military absolutism.

The unsavoury image of mercenaries, however, deepened with the mounting calls for political reform in France as King Louis XVI increasingly levied taxes to cover the cost of his ruinous military campaigns against the British:

> the successful war of independence waged by the United States may have assuaged somewhat the humiliations France had suffered from England in India, Canada and the Caribbean; however the war had cost over one billion *livres*, more than twice the usual annual revenue of the state
>
> (McPhee 2002: 35)

In 1789, foreigners made up about a quarter of the French Army: there were 12 foreign regiments out of the 91 non-Swiss regiments in the French line infantry. It is no wonder that the French military defeats and the financial burden of war were blamed on the mercenary troops in French employ.

Furthermore, when the Revolution broke out in May 1789, the Swiss and German regiments remained loyal to the king, exacerbating popular sentiment against their presence in France. Among its first decrees, the new Constituent Assembly called for the expulsion of foreigners from the French Army, which was finally achieved in July 1791. Ironically this transformation only occurred when German soldiers in the Nassau Regiment, after constant harassment from the French citizens, tore off their insignia and declared that they were French. The remaining Swiss units were largely massacred in 1792 and the survivors were subsequently disbanded.

The French Revolution encouraged the creation of a national army by pitting the citizens, with their grievances, against the sovereign and his army of foreigners. The consequences of this ad-hoc national force, however, were unexpected. Conscription was first introduced in 1790 with the aim of creating a homogenous army "based on the cheap supply of patriotic manpower, clearly superior to their more expensive predecessors in which loyalty to the organisation was low"

(Porch 2010: xvi). This citizen-army "expressed national purpose and fought for national goals, which made them, potentially at least, at once more forceful and more flexible (…). Nor were they as likely to threaten the integrity of the state" (ibid.). In reality, however, the overhaul of the armed forces was fraught with problems, political intrigues and mixed loyalties. Without its officer corps, which had largely resigned or been killed during the Terror, the army lacked discipline, training and strategy. By August 1792, after a disappointing performance against the Austrian cavalry and the capitulations of Longwy and Verdun, the government appealed to the Swiss to return, enlisting 3,000 to 4,000 Swiss mercenaries who had previously been released from service.

A second unexpected consequence of the French Revolution was its ideological appeal to soldiers and civilians across Europe who flocked into the country to embrace liberty, equality and fraternity. The propaganda machine worked too well, as France promised asylum to all victims of political repression, making it the '*terre promise*' for refugees, criminals and idealists. In August and September 1792, a *légion franche étrangère* and a *légion germanique* were created to integrate the excess Dutch, Belgian, Prussian and Austrian deserters into the French Army.

The First Republic thus began its days with a strong ideological appeal, but an inefficient citizen-army without proper experienced leadership, and an overload of foreign soldiers creating havoc inside the country. The solution to both these problems was to make an about-turn in policy and re-establish foreign regiments into the French armed forces. Napoléon Bonaparte maintained this tradition in his military campaigns across Europe, requiring the troops of conquered nations to serve in his army. Thus by the time Napoleon attacked Russia in 1812, a third to a half of his army was made up of former soldiers from Poland, Italy, Greece, Holland and Spain. These foreign regiments, however, were still considered inferior to French troops, and they were often assigned secondary roles, such as coastal defence and garrison duties. This precaution was justified in the high desertion rates of foreign soldiers, particularly in the Saxon, Bavarian and German units. By 1813, the government had dissolved most of its foreign units, with the exception of three *régiments étrangers* and the Swiss and Polish regiments.

Military historian Douglas Porch argues that the rise in nationalist sentiment that resulted from the Revolution actually gave a second wind, albeit arguably short-lived, to mercenarism. The ideological appeal and promises of freedom in France made the country into the homeland for mercenaries and "virtually guaranteed that generations of Europe's politically repressed would make for Paris".

The normative case against mercenaries

Despite, or perhaps because of their extensive role in the history of war, mercenaries have always maintained a bad image. They pillaged and plundered France during the Hundred Years War, wreaked havoc in the Italian city-states between the fourteenth and sixteenth centuries and pillaged central Europe in the Thirty

Years War. Their reputation as "faithless, dangerous and expensive" (Gothenburg in Paret 1986: 47) is well-earned. Nonetheless, mercenaries were useful and often more reliable combatants than citizens throughout most of history. They remained a legitimate feature of wars until states began to pass laws in the nineteenth century prohibiting their use and recruitment on national territory. This distaste for mercenaries, however, easily predates Niccolò Machiavelli's famous warning to Prince Lorenzo against the evils of mercenaries whom he qualified as "useless and dangerous (...), disunited, ambitious, without discipline, disloyal, overbearing among friends, cowardly among enemies; there is no fear of God, no loyalty to men" (Machiavelli 1992: 32). In *Barbarous Philosophers*, Christopher Coker interprets Machiavelli's disgust for mercenaries as recognition that these combatants no longer have a role to play in modern warfare. With the advent of new technology and the rise of civic-militarism, war experienced a transformation in the way that it was perceived and experienced and "in this paradigm shift, Machiavelli believed there could be no place for mercenaries" (Coker 2010: 143).

The general disapproval of mercenarism displays a normative judgement that is today institutionalised in international law. Norms are generally defined as "collective expectations for the proper behaviour of actors within a given identity" (Katzenstein 1996: 15). They reveal the shared values of a society and are constantly shifting to adapt to the cultural and technological changes that characterise modernity. Customary international law reflects the principle that laws are to be derived from the common behaviours of international actors. The laws of war, indeed, were a matter of customary law until they were codified first in the Hague Convention, then in the Geneva Convention. Hence it is only since the end of the nineteenth century that the general norm against mercenaries has made them into illegal combatants. Wars are normative and constantly changing, adapting to new methods, strategies, political systems and cultures and actors, among others. Consequently, norms regarding combatants have also fluctuated. This is true of the norms concerning the use of non-state actors, or mercenaries, in war.

Money, morals and military virtue

Mercenaries are perceived as immoral because they fight for money, have no attachment to the conflict and are unaccountable. This critical assessment of mercenaries is both a result of the terrible reputation that they have justly earned through their own behaviour in the past and relative to the ideal of the combatant that societies have strived to create through military institutionalisation.

Machiavelli envisaged an "Aristotelian figure of the armed and independent citizen willing to fight for his liberties" (Coker 2010: 145) as the model combatant and the only one capable of successfully waging war in conjunction with the norms of modern society. Because of their lethal potential and their necessity to the security of the state, warriors need to embrace restraint and civic virtue. This is the antithesis of mercenaries who are creatures of appetite in the Platonic sense, unable to control their murderous impulses and waging war for the sake of war itself, and for money. The uncontrollable and indiscriminate violence that has been associated with mercenaries since the Middle Ages only serves to confirm this accusation. Mercenaries are outside of the instrumental rationality of the state; they are therefore not controlled by the state and cannot be considered as part of the Trinity that characterise Plato's definition of the state.

Mercenaries today are considered morally wrong because they do not have any attachment to the cause for which they fight, and are therfore 'whoring' themselves by going to war 'just for money'. This goes against the norms of contemporary society: 'good' combatants should have an intrinsic interest in the outcome of the conflict. This intrinsic interest is anchored in nationalist sentiment, and therefore to a sense of belonging to a community for whose survival the combatant is ready to die. It guarantees loyalty and reliability. Anna Leander refers to the "intrinsically immoral nature of contractors" (Leander 2006: 76), which is derived from the idea "that to fight for money, without sharing the goals of the hiring group, is problematic" (Percy 2007: 244). Sarah Percy argues that mercenaries "are morally undesirable because they do not fight for an appropriate cause". Unlike citizen-soldiers, "it is their appetites, not reason of state, that propel them into war" (Coker 2010: 147).

Furthermore, through their alleged cowardice and lack of loyalty, mercenaries are an insult to military virtues:

> mercenaries are not considered to be good soldiers; they were businessmen only interested in profit; they did not take risks because they wanted to survive to old age to enjoy wealth, rather than die when still young on a battlefield for civic glory.
>
> (Ibid.: 143)

Contemporary society requires that their combatants follow an institutionalised set of rules that codifies their behaviour, rendering them into 'just warriors', reliable in war and loyal to the civilian community. Even Frederick the Great of Prussia, who made use of their services, criticised mercenaries for having "neither courage, nor loyalty, nor group spirit, nor spirit of sacrifice, not self-reliance" (Paret 1986: 108). Equally, King Gustavus Adolphus "took good care to keep them away from his native Sweden" (Porch 2010: xv) as they were both a danger to and a corruptive influence on society. Evidently, mercenaries do not fulfil this just warrior ideal.

Killing and control

The moral argument is extended to the idea of *jus ad bellum* "which define who can kill in warfare and separate out justifiable killing from unjustified murder" (Percy 2007: 54) and the *jus post bellum* which refers to the accountability of warring parties after a conflict. Modern states define themselves in terms of holding a "monopoly of the legitimate use of physical force within a given territory" (Weber 2004). By this norm, which is embedded in international law, only the state has the authority to sanction and delegate an act of violence within specified boundaries. It must therefore have the ability to control its combatants and hold them accountable for their behaviour during and after the war. This is not the case for mercenaries, however, as they are generally non-nationals operating outside of the territory of the state that hires them; they are beyond the reach of legal jurisdiction: "the international community's fear of mercenaries lies in that they are wholly independent from any constraints built into the nation-state system" (Percy 2007: 58). This problem is correlated to the foreignness of mercenaries and further justifies the norm against mercenarism.

Because mercenaries have no institutional, legal, or national ties to a conflict, they may also leave at any moment, making them not only unreliable but also uncontrollable. Unlike soldiers, mercenaries are outside of any juridical system of punishment, other than those that they set up themselves. In an unusual but extreme example, Colonel Mike Hoare, the commander of 5 Commando in Congo in 1964, ordered that one of his mercenaries have his big toe shot off as punishment for having raped and killed a young girl. Cesare Borgia, himself a *condottiero* and a duke in the fifteenth century, dealt with the mercenary problem in Sinigaglia (Italy) by inviting his three captains into his castle and subsequently having them strangled. Usually, though, mercenaries have been unaccountable for their actions, and this is still the case today. Despite proscriptive norms against mercenaries, neither governments nor the international community has been able or willing to control or sanction the behaviour of these non-state combatants.

Just cause and motivation

The normative case against mercenaries is very much entrenched in the accusation that these combatants have no justifiable cause for going to war. It also assumes that a financial motivation is not acceptable grounds either for killing or for taking such great risks with their own lives. Protocol I reflects this convention in Article 47: it defines a mercenary by his motivation "to take part in the hostilities essentially by the desire for private gain". The importance of the financial incentive is evident in the emphasis made on the difference in pay that characterises the earnings of a mercenary compared to that of a soldier: a mercenary "is promised, by or on behalf of a Party to the conflict, material compensation substantially in excess of that promised or paid to combatants of similar ranks and functions in the armed forces of that Party".

This focus on motivation, however, is the greatest legal weakness of the Geneva Conventions. Motivation is a state of mind, and while circumstances and bank statements can show proof that financial transactions were an incentive for participating in combat, it is virtually impossible to demonstrate that a combatant had *no other* personal, religious, or political motivations. This is evidenced in the 1972 Diplock Report commissioned in the United Kingdom to address the problem of terrorism in Northern Island, which concludes that any "definition relying on positive proof of motivation would 'either be unworkable or so haphazard in its application as between comparable individuals as to be unacceptable' " (Percy 2007).

Furthermore, the Diplock Report also argues that the motivations of non-state fighters range from "sheer desire for private gain accompanied by indifference to the cause which that force is supporting, to a conscientious conviction that the merits of the cause are so great as to justify sacrificing his own life". Although mercenaries generally *are* incentivised by the short-term prospects of earning a good salary, they are also motivated to join this trade by societal and economic circumstances, political conviction, or simply a passion for 'soldiering by other means'. This does not exclude them from being defined as mercenaries, in the sense that they are participating in a conflict in which society has decided that they have no intrinsic or moral interest, and are being remunerated for their efforts.

Profit and the economic realities of demilitarisation

The profit motive is often not sufficient to explain the career choice and huge risks that the mercenaries chose to take. The promise of riches can be exaggerated and is too frequently never fulfilled. Cognisant of this, Machiavelli explained that the "paltry pay" of mercenaries was not even enough motivation for them to risk their lives, making them cowardly and useless as combatants. Despite Machiavelli's accusation that mercenaries lack the courage to take risks and gamble their lives for the sake of money, the actual mortality rates for this profession suggest the opposite: in the Congo of the 1960s, the mortality rate for mercenaries went as high as "one out of four men in actual combat from death, wounds and disease" (Hoare 2008: 67). In Iraq and Afghanistan, more than 1,500 contractors have been killed, with a further 13,000 wounded between 2003 and 2007. Mercenaries, however, do not appear to associate their profession with the level of risks that they are taking: "that they might be killed in action was something that never entered their heads" (ibid.: 74).

Early in the war in Afghanistan, contractors were lured to the country by salaries as high as US$1,000 per day and were able to earn US$240,000 a year working in Iraq. Self-confessed mercenary Simon Mann makes a distinction between his motivation and those of the men fighting with him in Angola on behalf of Executive Outcomes and the oil companies:[5]

> these men were not doing this for their country. They were not doing this as Tony and I were – in order to defend their property and livelihood. They were being wounded, being killed in Willy's case – purely for money.
>
> (Mann 2011)

At the same time, in *Congo Warrior*, Mike Hoare described mercenaries' tendency to spend most of their pay while in the field, which suggests that making money and living a comfortable lifestyle were certainly not among their chief points of motivation:

what good is money to you in this god-forsaken part of the world anyway? (...) Tomorrow you may be killed in battle, with your horrible, hard-earned bucks still in your grubby little pockets. Lot of good it will do you then!

(Hoare 2008: 110)

To add insult to injury, mercenaries have often been victims of employers defaulting on contracts – occurrences which led French journalist Philippe Chapleau to refer to these men as "*soldats d'infortune*", or 'unlucky/unfortunate soldiers'. Indeed, to avoid paying these unconventional and potentially dangerous combatants, the princes and lords of feudal Italy repeatedly resorted to assassinating the leaders of the Condottieri. Thus such personalities as Carmagnola, Albert Sterz, Fra Moriale and Paolo Orsini saw their military career cut short by the very people who had hired them. Defaults on payments are still a significant risk in this sector. Tim Spicer, the CEO of the private military company Sandline, sued the government of Papua New Guinea for the US$18 million which the country owed him according to their signed contract. Despite early termination of the contract, Spicer won his case in 1999 and was eventually awarded the full sum he had been promised. Neal Ellis, a helicopter gunship pilot in Sierra Leone and one of the most famous mercenaries according to South African journalist Al Venter, claimed to be fighting because "flying over the jungle was better than working for a living". Venter explained that "this South African mercenary loved the job. He must have, because he certainly wasn't getting paid when I was there (...). By that time he and his crew were owed about a million dollars" (Venter 2006: 64).

The economic realities of demilitarisation, which left thousands of men without a job, have also been a launching ground for soldiers to become mercenaries. Rather than being motivated by profit, these men are just trying to survive and make a living as they are left unemployed once the guns fall quiet. The supply of mercenaries, therefore, rises at the end of wars in what Chapleau and Misser call a 'wave phenomenon':

after the wave of Vietnam, the English emerged on the market following the Falklands War; after the Gulf War, the offer was a bit more diversified: still the English, Americans, the French whose numbers went up after their intervention in Ex-Yugoslavia. More recently, many former soldiers from the Eastern bloc have flooded the market.

(Chapleau and Misser 1998: 85)

In particular, the end of the Cold War saw substantial military cuts and more than one million demobilised soldiers from the United States and Russia (Avant 2005). Some of these men developed skills in training for war that are not easily transferable in civilian society, and as a result, they remained unemployable after their military mandate expired. A South African Recces captain quoted by Chapleau asks quite justifiably "what else can you do, when you only know how to make war?" The demilitarisation of society, a result of changing political trends, has left many soldiers disappointed, disillusioned and questioning the integrity of their home state.

Lieutenant-Colonel Barlow, for example, explains that when the new South African government dismantled the infamous 32 Battalion, "these men, both black and white, who had readily sacrificed their lives for South Africa, were finally betrayed in 1993 by the country they had so willingly served and died for. We were deeply ashamed of this" (Barlow 2007: 129). Soldiers were left with a feeling of resentment and despair, and found it increasingly difficult to reintegrate into society. The lack of supporting measures to ease this transition exacerbated the situation. This resulting feeling of betrayal served to alienate a large proportion of the armed forces, many of whom became mercenaries as a way to continue exercising their profession, albeit no longer for their country.

Ideology

Although mercenaries are defined by a lack of political interest in the conflict in which they are participating, many freelance combatants have had or have developed ideological ties to the countries in which they are operating. Several white mercenaries fighting in Africa from the 1960s to the 1980s cited the communist threat as "one of their principal justifications for fighting the liberation movements which were provided with military aid, training and finances by both the Soviet Union and China" (Arnold 1999: 27). Eeben Barlow, for example, stresses that his company's operations in Africa were motivated by his desire to help "African governments that had been abandoned by the West and were facing threats from insurgencies, terrorism and organised crime. I believe that only Africans (Black and White) can truly solve Africa's problems" (Barlow blog 2008).

As a South African, Lieutenant-Colonel Barlow maintained strong ideological and political beliefs regarding the state of security on the continent, and had a vested interest, not just a financial motivation, to participate in the military processes of 'peace-building', particularly in Angola. In a comparable tone, Simon Mann expresses his sense of responsibility towards the repressed citizens of Equatorial Guinea as well as a desire for adventure and excitement: "I want to make the money. I want to make a difference, make some lives better. I feel challenged to takes on such a tough job. I want the danger and the hardship. I love the craic" (Mann 2011). Ideological attachment to a cause can develop with exposure, regardless of the circumstances that led the men to combat. The mercenaries fighting in Sierra Leone against the rebels became significantly attached to the difference that they could make by protecting the citizens from the brutality of the RUF. Journalist Al Venter insists that mercenaries have played an important role defending citizens and governments in countries that have fallen into war and chaos: these non-state combatants are "fighting other people's wars" (Venter 2011) and are stepping into the political vacuum that a non-interfering West has left post-Cold War.

A quest for adventure: the existential dimension

The risks that mercenaries take are not explained by economic gain. Machiavelli is right when he argues that men will not put their lives on the line just for the sake of money. Plenty of men joined the profession of mercenarism not only in pursuit of adventure but also in search of a sense of meaning which can arguably be reached through the hardships of war. Combat becomes "a supreme test of character in which those who come through achieve a lasting sense of self-knowledge of a kind usually not available in civilian life" (Coker 2007: 11). The empowerment of one's actions, the ability to exceed his capabilities, to transcend his own fears and contribute to a larger purpose within a group has motivated many a man (and woman) to pick up arms.

War has a deeply existential dimension that often redefines a combatant's sense of self-worth. This transformative experience makes the combatant *authentic* in the sense that he is only himself, his life is only worth living, when he is in battle. It is at this time that the individual is fulfilled in the adrenaline of the action and the ensuing recognition that he craves from others on the assumption that the sacrifice of his life is worth something to somebody. Men who derive this satisfaction from war "end up in armies and many more move on again to become mercenaries because regular army life in peacetime is too routine and boring" (Dyer in Grossman 1995: 180).

Mercenaries who have been drawn to Africa since the 1960s with the promise of adventure have lauded this new avenue for professional combatants:

> the mercenary unit of soldiers in these days is unique, so you could say it is the opportunity to practice our skills which attracts us rather than the way of life. (...) The ambitions of a soldier can be consummated only by action.
>
> (Hoare 2008: 141)

The fact that so many mercenaries (but not all) are indeed former members of their nation's armed forces with combat experience also shows, at least to a certain extent, the attachment that these men have to the life of soldiering. On the other hand, some men are just attracted by a license to kill, as well as by the state of chaos that inevitably results from conflict and leads to opportunities for crime. In the Rhodesian war, a number of mercenaries were allegedly "paid by the kill"[6] and "a few were there just for the love of killing" (Venter 2011: 9).

In their forays into the world of private contractors, journalists Robert Young Pelton and Steve Fainaru found that, although these men were motivated by patriotism and money, they also shared an intrinsic interest, not to say love, for this pseudo-military lifestyle. A Marine Forward Recon working for DynCorp and interviewed by Pelton revealed that "he saw working the Karzai detail as a way to make money, continue his interest in the military, and be part of a unique moment in history" (Pelton 2007: 77). Some contractors "develop an addiction to the lifestyle and a dark craving for being 'in the game'" (ibid.: 340). Fainaru encountered these characters in Iraq, as he was imbedded with the cost-cutting, corrupt private security company Crescent Security Group: former Marine John Munns explained his choice of career by his "need (for) something to show my system to remind myself I'm alive" (Fainaru 2008: xiii). Another young contractor, Jon Coté, who was later kidnapped and assassinated in Iraq, stated that his job with Crescent was "by far the coolest thing I have every experienced in my life" (ibid.: 2). Fainaru concluded that Iraq and the private security industry ensnared men with its promise for "the camaraderie and the addictive thrill – Iraq as a reality, not as an abstraction" (Fainaru 2008: xvi). Each contractor also had their personal psychological motivations and perversions that led them away from the safety of their homes and into a war zone: "you were part of it, and it was part of history, and so you were part of history too, even if you were dead. But it went much deeper, and it was mostly personal" (ibid.: xvi).

Conclusion: mercenaries as non-state warriors

Mercenaries can exhibit the same military virtues as soldiers according to Deane-Peter Baker. He argues in his book *Just Warriors, Inc.* that "there are no intrinsic flaws which apply generally to the character of private warriors such that they are unable to display courage, exhibit comradeship and a sense of discipline, or lack an appropriate sense of honour" (Baker 2011: 63). Aristotle offers a different valuation of courage, claiming that men who are brave and virtuous are happier than others but have more to lose by dying:

> for life is best worth living for such a man, and he is knowingly losing the greatest goods, and this is painful. But he is nonetheless brave, and perhaps all the more so, because he chooses noble deeds of war at that cost.
>
> (Aristotle 2009)

The best soldier, therefore, "may be not men of this sort but those who are less brave but have no other good; for these are ready to face danger, and they sell their life for trifling gains" (ibid.). According to Aristotle, the *type* of courage exhibited by a soldier and a mercenary differs in its ultimate purpose.

The transcendental experience of war is not equivalent for mercenaries as it is for soldiers. Both combatants can derive a sense of self-worth and existential satisfaction from their role in combat. Only soldiers, however, are considered as worthy of honour and recognition. Whereas society values the sacrifice of soldiers, it vilifies the mercenary. This is significant, because it is the recognition of the combatant's sacrifice that gives value to the life and death of the warrior. The importance of recognition cannot be undervalued in the experience of the combatant. Coker describes the warrior as an instrument of the state, who derives his purpose from his desire for glory, vainglory, *amour propre*, and recognition: "man is the only animal who needs to place value on things – himself in the first instance but also on people and actions around him, such as the unit, or regiment, or 'band of brothers'" (Coker 2007: 14). Therefore, Shannon French explains, warriors are not mere tools (…). For those who send them off to war must make an effort to ensure that the warriors themselves fully understand the purpose of, and need for, their sacrifice. Those heading into harm's way must be given sincere assurances that their lives will not be squandered.

(French 2003: 10)

Mercenarism does not cater to the 'warrior experience' of combatants who derive their sense of self-worth from the recognition of their societies. This is unavoidable, since mercenaries isolate themselves from their own homes by agreeing to fight for the cause of another country, an act that is not normatively accepted, even in this age. In the process of executing their contractual arrangement, they lose their sense of belonging and forfeit their right to recognition.

One of the main differences between soldiers and mercenaries lies in the value that society puts on the exercise of restraint. Soldiers operate within a legal framework and are furthermore indoctrinated in a warrior code that defines how they should fight, who they should kill, when, and how. The *why* is mandated by society. Shannon French suggests that the warrior code is set up "to protect the warrior himself from psychological damage" in the face of the murderous task that he/she is set in the process of defending his/her country. The warrior code also serves to create ties between the men and women who share these principles and increases their effectiveness in combat by setting common standards of behaviour which reinforces their mutual trust and cohesiveness.

The principle of restraint is the guiding factor behind the laws of war and the Geneva Convention. Restraint is necessary to ensure that prisoners of war will be treated humanely and not murdered. It protects the civilians from arbitrary retribution. In contemporary strategies of counterinsurgency, restraint is a central element of the hearts and minds campaign. It also marks the distinction between murder and lawful killing and only state-sanctioned killing is deemed honourable. The importance of restraint as a warrior virtue is most vivid in Shannon French's survey of military personnel, who value "a Hector who wins" over the legendary Achilles in Homer's epic poem. Although Achilles is required to go to war to defend King Menelaus' honour, he has no real interest in the battle and "does not really care about the Greek cause or the condition of the Greek troops" (French 2003: 43). His behaviour in combat shows no restraint or respect for his enemy. He has no warrior code but instead "sets his own standards of conduct and relies exclusively on his own internal judgements of when he is deserving of honour and when he should feel shame" (ibid.: 60). The opinion of his peers is meaningless to him and his final desecration of Hector's body shows an utter lack of restraint.

This is contrasted to Hector whose "humanity and nobility of character" are admired by contemporaries. The Trojan prince derives his sense of self-worth from his position as the defender of his city: "he cannot now, in the midst of such a dreadful conflict, allow himself to be anything other than Troy's bravest defender without losing his self-respect and all sense of identity, meaning, and purpose in his life" (French 2003: 33). Although both warriors are celebrated in history for their military prowess, they represent opposite values in their experience of war. After Hector's death, Achilles is left as "a warrior with no cause he truly embraces, far from his own country, with no beloved comrades left to fight for or revenge" (ibid.: 51). The only thing that separated him from being a mercenary is that, despite his lack of restrain, Achilles' actions were entirely sanctioned by his kinsmen.

Today, soldiers are expected to exercise restraint. Killing must still be state-sanctioned, but the behaviour of combatants on the field is strictly codified by the Geneva Convention. Achilles' behaviour would be punishable within the context of the army in which he belonged, whereas mercenaries lie beyond the realm of accountability. Despite any 'warrior' or moral code that they may exercise, regardless of their actual motivations or behaviour, mercenaries are not considered to be warriors. They are legally unaccountable and uncontrollable to a society that values sacrifice, restraint, and control above all warrior virtues. In the eyes of the international community, mercenaries are illegal combatants, undeserving of recognition, unrestrained by laws, and unattached to any legitimate cause.

Notes

1 This decisive battle took place in the territory of Iraq.
2 Mercenaries in the Carthaginian Army included "Celtes, Gaulois, Ibères, Cantabres, Baléares, Lusitaniens, fugitifs de Romes, Ligures, cavaliers tarentins, Doriens, Grecs, Ioniens, Lacedemoniens, Lydiens, archers de Cappadoce, Lybiens, Egyptiens, "Nègres" et Indiens" – Philippe Chapleau and François Misser, *Mercenaires S.A*, Paris: Desclée de Brouwer, 1998: 15.
3 Ironically, Hamilcar's son, Hannibal, became the military strategist who successfully led his army of African mercenaries and European tribesmen against the more numerous Roman citizen-army at the Battle of Cannae in 216 BC.
4 The last war that Switzerland fought in was the Sonderbund War in 1847 when the Protestant cantons defeated the Roman Catholic cantons that had formed the separatist Sonderbund league.
5 Mann's claim to be fighting for his livelihood on the basis of selected investments in Angola is arguably no different than the motivation of the mercenaries he berates.
6 Interview with a South African activist, Cape Town, 15 March 2011.

3   Soldiers and national security

Soldiers are agents of the state: recruited, paid, trained, punished and sometimes killed by their employer. Their citizenship makes them liable for military service, responsible for the defence of the nation and accountable to the state. Unlike mercenaries, armies of soldiers represent the state's preference for total control over its military agents: armies are the "operational tools of rulers interested in creating stable social relationships based on a monopoly of violence" (Glete 2002: 5). The soldier's enforced subservience in war, ruthless military training and national indoctrination within the context of the army inevitably leads to a different state of mind than that of the mercenary. Consequently, mercenaries and soldiers view war and duty through opposite lenses, experiencing the same horror of combat but with very different responsibilities framing their actions and relationships.

This chapter presents the soldier's rise to prominence as the chosen combatant of the state. It describes, and dispels, the specific conditions under which soldiers are expected to develop a sense of patriotism and duty for which they are ready to die – a motivation which mercenaries allegedly lack. As an agent of the state, the soldier loses the right to autonomy, which is arguably at the origin of his resentment and distrust of mercenaries: legally and psychologically free to leave when the situation becomes too dangerous, mercenaries make unreliable allies and are thus a direct threat to the lives of the soldiers next to whom they are meant to be fighting in select conflicts.

Definition of a soldier

The first attempt to regularise armies came in 1445 when King Charles VII of France created the *Compagnies d'Ordonnance* that hired mercenaries on a permanent basis to defend his kingdom. This offered the realm an unprecedented capable and reliable force, led by senior officers appointed by the crown and loyal to the monarchy. These companies were dependent on the king for their supplies and salaries. The first men to receive these regular salaries were called *soldats*. The word *soldat* comes from the Italian word for money, itself derived from the Latin gold coin called *solidus*, and means literally, one who receives the solidi – the money or salary. Indeed, John Keegan and Richard Holmes define soldiers as "warriors who fight for pay" (Keegan *et al.* 1985).

A soldier today is defined as a combatant who is an official member of a permanent army sanctioned by and under the jurisdiction of the nation-state. There are several types of combatants who fall under this category of 'national soldiers': contractual volunteers who have chosen to enrol in the armed forces for a limited amount of time; career soldiers, generally officers and non-commissioned officers (NCOs); conscripted soldiers, who are expected to perform a civic duty towards their state; and mobilised soldiers, who are called upon in times of crises to defend the country. The common denominator between the different types of soldiers is their dependence, subservience and accountability to the state, whose nationality they share, during the entire duration of their contractual agreement.

The birth of the citizen-army

Citizen-armies can be organised into three main categories: Aristotle and Plato conceived of the first citizen-soldiers in the context of Ancient Greece, with the polis as the principle political actor. Machiavelli's concept of a citizen-army in the city-state of Florence took on a different form in a diversified geopolitical environment which was described in the previous chapter in the case study of the White Company. Finally, the eighteenth-century idea of the citizen-soldier emerged from Jean-Jacques Rousseau's model of the social contract in which all citizens entered a covenant with the state in which they agreed to participate in the common defence of their society in exchange for the protection they enjoyed. These categories of citizen-armies reflect the changing norms of 'civil society' as defined by Adam Ferguson: in the first, armies fought princely wars, in the second, armies fought for profit, and in the third, they fought for the virtue of the state.

The concept of the citizen-soldier was first articulated by Aristotle in the *Nicomachean Ethics* in which he explains that virtue and citizenship are the key attributes of the *polis*. In exchange for the privileges of being a citizen in a city or state, men were expected to contribute to the protection of their society by taking up arms whenever necessary, and without expecting financial compensation. On the other hand, sailors, known as *nautus ochulos*, were pooled from the poorest citizens of Athens and therefore were paid for their work. Plato dismissed these men because he considered them demeaning to the integrity of the state.

Civic duty in the Middle Ages was extended to the feudal system whereby serfs owed their rulers 40 days of military service per year. Large armies of citizens, however, only became the norm in the nineteenth century. Prior to this, rulers mostly fought their wars through non-state subsidiaries that were hired for the duration of the conflict and subsequently released from their duties. Rulers were reluctant to arm their citizens for security reasons, and the general population was "eager to let any available foreigners assist them in any necessary bleeding and dying for *la Patrie*" (Porch 2010: xi).

Early attempts by Machiavelli to field an army of citizen-soldiers, which he considered more efficient and trustworthy, in the city of Florence failed as the

Florentine militia was defeated by the Spanish regulars in 1512, leading to the dissolution of the city-state and Machiavelli's prompt dismissal and imprisonment. Instead, mercenaries continued to be the norm up until the French Revolution and the Revolutionary Wars: the Thirty Years War that brought about the Treaty of Westphalia – and sowed the first grains of sovereignty – was largely fought by mercenaries who were led by captains with personal allegiances to the state. France was dependent on foreigners for most of her wars: "from the Scots who rode with Joan of Arc to the Foreign Legion and Dien Bien Phu, the foreign soldier, idealistic volunteer or hard-case mercenary, is an integral part of the French military tradition" (Elting 1988: 135). Britain has equally used foreign and proxy troops in her colonial projects up to and including the American War of Independence and during the Crimean War. It was only after the French Revolution that ideals such as patriotism and civic duty spread across Europe and eventually led to the adoption of conscription and the creation of citizen-armies.

Citizens, enlightenment and the Revolutionary Wars

The ideas that emerged from the French Revolution changed the military fabric in Europe for at last the next 200 years. The Enlightenment that preceded the Revolution proved to be a cultural phenomenon. Concepts such as the social contract and citizen rights were particularly appealing in France where the monarchy and nobility abused their power of taxation, enforcing excess duress on a population already suffering from the 1787 financial crisis and several years of bad harvest. Information and propaganda were communicated through pamphlets and publications that were increasingly accessible thanks to the availability of new technology: "well before 1789, the language of 'citizen', 'nation', 'social contract' and 'general will' was articulated across French society, clashing with an older discourse of 'orders', 'estates', and 'corporations'" (McPhee 2002: 31).

The spontaneous uprisings that spread across the country leading up to the assault of the Bastille prison reflected a population's exasperation with an antiquated socio-political system that was unable or unwilling to adapt to the requirements of a new era. When the French Revolution broke out in 1789, the French Army was largely made up of foreigners and noblemen whose duty it was to protect the royal family and execute the king's decree upon his people. Understandably, the population's hatred for the monarch turned against his military representatives: the nobility and the foreigners charged with his protection. The call for a system of conscription was finally realised in 1793 as an alternative to foreign regiments. Known as the *levée en masse*, the first draft initially ordered 300,000 men into the armed forces, followed by the general mobilisation of all young men, an event unheard of in military history.

The chaos of the Revolution, purging of officers and disbanding of all foreign troops had created a military vacuum by 1792. Furthermore, the citizen's uprising against the monarchy insulted the neighbouring royal families who subsequently issued the Declaration of Pillnitz, a statement of support for the French monarch that threatened retaliation if anything were to happen to him. This, however, only served to further inflame the citizens of France who declared war on Austria and on Prussia in the spring of 1792. Inspired by ideological discourse and nationalistic sentiment, Frenchmen flocked to the army, overwhelming the smaller Prussian, Austrian and Dutch armies with their numbers. Between 1789 and the Peace Treaty of Amiens in 1802, the French Republic with its army of conscripts had defeated all its enemies and conquered neighbouring territories that had eluded the ambitions of the Valois and Bourbon monarchs.

Clausewitz attributed French successes in the Revolutionary Wars to the enthusiasm of its citizen-army. Donald Porch and Tim Blanning, however, argue that the failures of an undisciplined and paranoid army led to France's early defeats in the Revolutionary Wars: in April 1792, the French Army "broke and ran at the first sign of enemy cavalry" (Porch 2010: xvi) and lost Longwy and Verdun to an allied army of Prussians, Austrians and Hessians. The disconcerting rate of desertions by French citizen-soldiers forced the Republican government to re-enlist Swiss troops and integrate other voluntary foreigners into their ranks. It was only after the reorganisation of the French Army and the reintegration of foreign mercenaries that the tides of war began to change and France started collecting victories on the battlefield.

Napoleon's Grande Armée and military reform

Contrary to popular belief, Napoleon's *Grande Armée* was not made up of revolutionary citizens imbued with enthusiasm and nationalism. Despite the *levée en masse*, by 1810, 80 per cent of conscripted men failed to report for duty. Instead, Napoleon relied heavily on foreigners who were forcibly incorporated into the French Army.

Napoleon's charismatic and intelligent leadership set the scene for the armies of the next century. Although most of his soldiers were poorly trained and equipped, Napoleon nurtured the formation of the educated military elite through the establishment of two officers' schools: Saint-Cyr in 1802 and the Ecole Polytechnique, which formed officers and promoted scientific research with the motto *"ils s'instruisent pour vaincre"* (they learn to win). Few officers from Saint-Cyr, however, actually fought in the Napoleonic Wars as these ended in 1815. In fact, most of Napoleon's generals were veterans of the *ancien régime*. Napoleon also developed a meritocratic system by which commissions were earned, and not bought, thus placing competent officers at the head of the army.

Most significant for Tim Blanning, however, is the emperor's ability to mobilise such a huge body of men on the battlefield. Before the Napoleonic Wars, most armies were of moderate size, with few exceeding 200,000 soldiers. Even the Prussian Army in which Clausewitz fought never exceeded 320,000 men at arms in the eighteenth century. By contrast, the *Grande Armée* amassed up to three million Frenchmen out of a population of 28 million. This was enabled partly by a mandatory military service for all males between the ages of 20 and 25. Blanning, however, successfully dispels the belief that the French

Army was victorious because of its energy and enthusiasm, and instead demonstrates that "whenever the allies were able to assemble even roughly the same number of troops as the French, they won" (Blanning 1996: 641). This shows that the French Army was not necessarily better than any other European army, but instead was able to dominate its opponents on the battlefield simply by overwhelming them with numbers that were reached by conscripting all available manpower from France and its conquered territories.

Napoleon's strategy of total war further contributed to the war effort as "until total victory was achieved, every man, woman, child, animal and inanimate object was conscripted for the war effort" (ibid.). Total war marked a dramatic change from the wars of the *ancien régime* which, although "an indescribable bloody horror" (Avrom 2007: 44), were kept on a leash and fought by noblemen who preferred to withdraw their best troops than sacrifice them to the enemy. Napoleon's armies were able to beat the Prussians because the French fought with total abandon: "the absolute destruction of the enemy became a moral imperative" (Bell 2008) whereas the demographically inferior Prussian officers were still using the old, cautious rules of war.

The changing military landscape of Europe

The military successes of Napoleon's army changed the military landscape of Europe for two reasons: first, Napoleon's conquests in Spain, Italy and Prussia forged a national identity within these territories that previously had never needed to exist. The population of the conquered territory had a common enemy in the occupying French Army. They were also united for the first time under one ruler. Inspired by the philosophy of the French Revolution, combined with the unwelcome experience of French occupation, the Italian city-states began a movement of insurrection against Napoleon, and later against the Austrian and Habsburg empires. Indirectly, the French conquest of Italy arguably contributed to its unification fifty years later. A similar case can be made for German unification. The occupation of the Rhineland and the enforced Frenchification of Germany contributed to the rise of nationalist sentiments that emerged from the populations' efforts to defend their traditions and values.

Second, the overwhelming numbers of the French Army and its use of total war set a new standard for European armies that wished to compete on equal footing with the French: "the French had introduced universal military service in 1793 via the *levée en masse*, inaugurating the era of mass continental land armies. The other European powers had to follow suit or risk their armies being vastly outnumbered" (Moore 2009: 37). Governments could no longer afford to field armies of mercenaries, and were forced to look to their own citizens for cheap military labour: "once the citizen-army was adopted by one state, the logic of path dependency took hold" (Percy 2007).

Having been defeated and occupied by the French, Prussia took the lead in military reforms, founding new officer training schools, opening up the officer corps to all classes of population, tying promotion to a system of examinations, and setting up a system of recruitment called the 'Krümper system' which would eventually turn into universal conscription. The Prussians, however, did not have a common spirit of patriotism after the Napoleonic Wars, and neither did the French or the British associate nationalism with a sense of military duty: "when Prussia adopted a citizen-army, her leaders were taking an extraordinary leap of faith. They had to hope that a standing army would create nationalism in a people notably lacking in it, and that nationalism would lead to better and more unified soldiers" (Percy 2007). The Prussian middle class, however, were not allowed to join the army because most of them were traders, and therefore not considered to be good fighters.

The introduction of conscription, therefore, created a new relationship between the state and its citizens. Rousseau's concept of a social contract was developed and exploited by governments to instil a sentiment of duty and patriotism into their military institutions: "now that the government was responsible to the people, the soldier was their servant – no longer an enemy but an ally" (Hunt 2012). George Washington expressed this relationship in terms of individual and communal responsibility on the part of the citizens towards the security of the state:

> it must be laid down as a primary position and the basis of our (democratic) system, that every citizen who enjoys the protection of a free Government owes not only a proportion of his property, but even his personal service to the defence of it.
>
> (Moore 2009: 76)

Percy argues that the ideals of the French Revolution and the Enlightenment, along with the military successes of citizen-armies, led to a new norm in European societies in which, for soldiers to be respectable, they had to voluntarily contribute to the war effort out of love for the nation.

By 1871, the militarily superior and reformed Prussian Army had crushed the Austrian, Danish and French armies in the span of just seven years. The French, having practised a form of modified conscription since the fall of Napoleon, were shocked and humiliated by this defeat. Their reaction was a massive military expansion in the following years. The armies that met one another on the battlefields of the First World War were unprecedented in their size and make-up:

> any continental power which wished to escape annihilation as swift and overwhelming as that which overtook the Second Empire had to imitate the German pattern and create a Nation in Arms – a nation whose entire manpower was not only trained as soldiers, but could be mobilized, armed and concentrated on the frontiers within a few days.
>
> (Howard 2001: 455)

Between 1914 and 1918, 60 million soldiers were mobilised across Europe, with governments appealing to patriotism and nationalist sentiment, and using conscription to forcibly recruit the numbers that were needed to face the enemy. The French Revolution and the subsequent wars between increasingly large European armies changed the military landscape of the continent. The systematic creation of citizen-armies reflected an instrumental belief in the importance of loyalty and motivation to defend the nation:

> the armed forces became an institution for solidifying the attachment to the nation and the state, for promoting social mobility, for education and thus played a central part of the shaping of the modern state, not only militarily but also socially and politically.
>
> (Leander 2006: 41)

The soldier and the state

The rise of massive citizen-armies in the nineteenth century placed a new burden on the state. To mobilise its citizens for war and turn them into reliable soldiers, the state needed to develop and instrumentalise three sentiments: nationalism, patriotism and a sense of duty. This was achieved militarily through the introduction of universal conscription, which reduced the social barriers and ensured that the masses were represented throughout the military ranks. Conscripts were generally expected to be literate and were educated in the glorious history of their nation as part of their military training. The state also began to invest heavily in the construction of new schools whose curricula were undeniably patriotic. Historian Eugen Weber explains that new educational policies made schools the main driver of patriotism and contributed significantly to the development of a homogenous nation-state.

In *All Quiet on the Western Front*, protagonist Paul Bäumer and his schoolmates enlist in the armed forces after being urged to do so by their school teacher. The state also encouraged a wave of propaganda that was "disseminated in the army camps in the forms of pamphlets and journals" (Posen 1993); they staged public displays of punishment for acts of cowardice, and ceremonies rewarding acts of heroism. In France, festivals celebrating the ideals of the Revolution brought together soldiers and civilians whereas patriotic songs were commissioned and taught in the schools. The entire country was mobilised in a cultural campaign to instil pride and patriotism among the citizens and ensure its emotional commitment to the security of the state.

Civic-militarism, duty and sacrifice

William Doyle explains that inherent in the ideals of the post-Revolutionary state is the notion of civic-militarism. This concept was developed by the Ancient Greeks for whom all free citizens of the state were expected to "fight to protect the commonwealth which in the exchange had granted them rights" (Hanson 2001: 123). Civic-militarism is articulated in the philosophy of the social contract which views the relationship between the citizens and the state in terms of dues and duties *(des droits et des devoirs)*. In the *Declaration des Droits de l'Homme et du Citoyen*, the right to electoral and judicial participation is expressed in Article VI: *La Loi est l'expression de la volonté générale. Tous les Citoyens ont droit de concourir personnellement, ou par leurs Représentants, à sa formation* (the law is the expression of the general will. All citizens have the right to compete either personally or through their representatives). This right, however, is conditional on the security of the polity, which requires the general participation and contribution of each individual: Article XII: *La garantie des droits de l'Homme et du Citoyen nécessite une force publique* (the guarantee of the rights of man and the citizen requires a public force). Article XIII therefore articulates the duty of military participation that is expected from each and every citizen: *Pour l'entretien de la force publique, et pour les dépenses d'administration, une contribution commune est indispensable. Elle doit être également répartie entre tous les citoyens, en raison de leurs facultés* (for the maintenance of a public force and for all the administrative expenses, a common contribution is indispensable. This burden must be shared equally by all citizens according to their abilities). This military duty, therefore, "is instrumental: it is in the service of others" (Coker 2007: 6), but it is also inescapable as it is tied from birth to the individual's right to exist, live and work inside the confines of the state.

Soldiers, therefore, are citizens who are actively fulfilling their duty towards the state. They are required to commit a portion of their time and potentially sacrifice their lives to protect the existence of their fellow citizens. It is this sacrifice that also explains the state's choice of citizen-soldiers over mercenaries. The Greek philosopher Aristotle compared citizen-soldiers to mercenaries whose training and experience, he admitted, make them "most capable in attack and in defence, since they can use their arms" (Aristotle 2009). Whereas mercenaries can be brave in battle, however, their bravery is not one which holds them to their post when faced with certain death: they "turn cowards when the danger puts too great a strain on them and they are inferior in numbers and equipment; for they are the first to fly, while citizen-forces die at their posts" (ibid.), as evidenced by the Spartan citizen-soldiers who stood their ground to the death in the famous Battle of Thermopylae in 480 BC. Aristotle explained that only citizen-soldiers have sufficient interest in the outcome of the battle to face the enemy with the necessary courage to stand their ground: "the courage of the citizen-soldier is most like true courage" (ibid.) because the individual is motivated by a sense of duty which drives him to avoid disgrace, pursue honour and sacrifice himself for his country. This notion is supported by Coker who states that each soldier has a personal responsibility for and intrinsic interest in the outcome of the battle which makes him a more dependable combatant. Furthermore, most soldiers in conscript armies were originally peasants and farmers, and "farmers don't yield ground" – an attitude that the wandering mercenaries evidently do not share.

Plato expands on the importance of duty in *The Republic*, where he defines the characteristics of soldiers, whom he calls Auxiliaries: "nothing can be more important than the work of a soldier should be well done" and the existence of these Auxiliaries therefore is instrumental in providing the security of the state. The soldier must be engaged to carry out his duty to the polity to the extent of sacrificing his own life. To Plato, Auxiliaries must be professionals whose bodies and soul are educated in the art of war and in the values of the city which they promise to protect, and to which they belong. They are characterised by their 'spirit' which stimulates their emotions and drives their actions to defend the parent-state: "this ferocity only comes from spirit, which, if rightly educated, would give courage, but, if too much intensified, is liable to become hard and brutal".

Citizen-armies re-emerged in the nineteenth century along with a new moral norm among states by which citizen-soldiers were perceived as the only appropriate combatants to represent their state on the battlefield: "if citizens were willing to die for their state, it suggested that the state was a powerful and glorious entity that took care of its people, who returned the favour" (Percy 2007). Patriotism and nationalism came to define the moral value of the state, and was expressed through the sacrifice of its citizens. Soldiers, therefore, became the preferred combatants because they fight for the right reasons. Having entered a covenant with the state by choice or by default of being a citizen of the state, they "serve a larger human purpose; (the soldier) puts himself at the disposal of his city, to enhance its power or secure its ends" (Coker 2010: 184).

Obedience and punishment

Despite the state's expectation that its soldiers are willing participants in the military project of the nation, it has nonetheless developed a system to guarantee the patriotic sacrifice of its servants. This is done through three measures: (1) draconian training with the objective of instilling blind obedience; (2) societal norms that glorify this obedience and vilify dishonour; and (3) military and civilian courts to punish deviants and discourage defiance. This infrastructure is necessary to ensure a soldier's performance on the battlefield and to avoid an abuse of military power that may threaten the civilian government: Plato warned of the dangers of the soldiers, who like Achilles, fall victim to the temptations of their spirit, and argued that only the state structure keeps these combatants under control. This is why Plato stressed the characteristics and training that are essential to the warrior class: "it is precisely because he has it in his power to pressure his fellow citizens that he needs to exercise remarkable self-restraint" (ibid.: 186). Plato explained that soldiers are not to have

> any property of his own beyond what is absolutely necessary (…). Their provisions should be only such as are required by trained warriors, who are men of temperance and courage; they should agree to receive from the citizens a fixed rate of pay, enough to meet the expenses of the year and no more; and they will go to mess and live together like soldiers in a camp.

Plato argued that by isolating the soldiers and removing any temptation of physical belongings and comfort, the Auxiliaries would devote their entire existence to protecting the polity to which they belong, and would not be tormented by their own 'appetites' which might lead them to irrational and violent behaviour.

Armies distinguish themselves from civilian institutions by their cohesive and hierarchical structure. Samuel Finer describes the army's main characteristics as having "a centralised command, hierarchy, discipline, intercommunications, esprit de corps and a corresponding isolation and self-sufficiency" (Finer 1962: 6). As their principal objective is to "fight and win wars", or to act as a "protection device against external enemies" (Feaver 2003: 8), the army differs in both function and purpose from civilian society. Soldiers end up thinking in terms of 'us' and 'them', where civilians and their institutions are perceived as inferior to the strictly disciplined life of the military.

The entire training mechanism of the soldier is therefore manufactured to forge him into a combatant and prepare him to follow the orders of his superiors. According to Clausewitz, "the end for which a soldier is recruited, clothed, armed and trained, the whole object of his sleeping, eating, drinking and marking is *simply that he should fight at the right place and the right time"*(Clausewitz 2006: 95). Basic training begins with "the socialisation and indoctrination of the recruit" by breaking down a man's civic identity and rebuilding a soldier in the image of the army (Moore 2009: 75). This psychological transformation is accomplished through a purposeful segregation of society in terms of 'military' and 'civilians'. Soldiers are distinguished by a common uniform and have the right to bear arms. They are kept apart from civilians with military barracks isolating the armed forces from the general population. Recruits are indoctrinated with the virtues of the military system with an emphasis on the unique qualities that only the armed forces embody. This includes their access to and monopoly over the use of arms, their organisational superiority, and their important symbolic status within the state. Humiliation and brutality are also used "to break down an individual's self-esteem, lower their resistance to the values and attitudes that the military wants them to adopt, and reinforce the omnipotent nature of military discipline" (ibid.).

Furthermore, the absolute regulation of the recruit's life inside the army base restricts his freedom and increases his dependence on the military institution. The recruit is told when to eat, sleep and fight. Personal initiative is discouraged and instead he is taught to act only upon hearing his superior's orders. A soldier's identity is stripped away and replaced with a rank and a function. Eventually, the recruit is broken down by the harsh physical training: sleep deprivation, long marches, repetition of drills, shooting and manoeuvres gradually condition the troops to obey without reflection to the verbal commands of their officers. Studies have shown that a rigorous training increases the soldier's commitment to the group which in turn improves his ability as a combatant. Rather than inspire mutiny, French legionnaire Captain Morin de la Haye explains that the hard training and education of the soldier produce surprising results:

one would say that the constant exercise of the will of the leader works on them like a hypnotic suggestion. One sees in the eyes of the soldiers that they are attentive to orders, proud to manoeuvre well, and conscious of their worth. (The instructors) carry it out with zeal, even fanaticism, in the expectation of campaigning with the men that they prepared for this end.

(Richard 1890: 132)

The military's strict hierarchical structure also encourages a culture of obedience. Soldiers are organised by rank in an order of increasing authority and responsibility. Refusal to accept orders is insubordination and is punishable within the military legal system. In the United States, for example, Articles 90 to 92 of the Uniform Code of Military Justice make it a crime to disobey the orders of a superior commissioned or non-commissioned officer. The exercise of hierarchy in the army is enabled by a legal body that is specifically created for the armed forces. By joining the military, the recruit forfeits his civilian rights: unlike other citizens, soldiers are governed by a specific body of laws that restricts their personal freedoms and rights; *"les militaires jouissent de tous les droits et libertés reconnus aux citoyens. Toutefois, l'exercice de certains d'entre eux est soit interdit, soit restreint dans les conditions fixées par la présente loi"* (all soldiers enjoy the same right and liberties given to citizens. At the same time, the exercise of some of these are either forbidden or restrained within the conditions described in the present law). These laws also cover the political life of the soldier who is forbidden from being politically active, syndicalise or go on strike in certain countries. Furthermore, while serving his country, the soldier is required to wear a uniform, live where he is told, and may only leave the military base with the explicit permission of his superior. In France, the *Code de la Défense* describes the legal requirements of the soldier, which range from the expectation of a spirit of sacrifice to the details regarding his availability, loyalty, neutrality and discipline. Its equivalent in the United States is the Uniform Code of Military Justice which regulates the disciplinary measures that can be taken against a member of the Uniformed Services.

Military bodies, therefore, are empowered by the civilian government with the right to punish any member of the armed forces according to the country's code of military justice. Offenders may undergo a military trial – as in the case of the United States court-martial system, and can be incarcerated in a special military prison or detention centre. Should a soldier fail to obey military orders, he is liable for punishments ranging from monetary fines, floggings or imprisonments to torture and death depending on the crime and the military culture. In the Roman Army, for example, treason was punished by placing the perpetrator in a bag of snakes and throwing him into a river or lake. Decimation, or the removal (by stoning or clubbing) of one in ten soldiers, was also used to punish cowardice and mutiny. This was particularly effective, because soldiers organised in these units of 120 men were likely to at least know each other by name and would therefore be personally affected by the execution of their comrades. Execution by firing squad can still be expected in many contemporary armies dealing with cases of high treason or spying. The entire military infrastructure is built around a system of strict juridical control, engrained with the values of the nation, and judged upon the normative preferences of the society.

Honour and recognition

Soldiers have neither the right nor the ability to refuse to partake in warfare: "the disciplinary apparatus of the military removes a soldier's ability to refuse a mission" (Moore 2009: 15). Soldiers are not expected to support the war – they are required to follow their orders as dictated by the contract between them and their state. In return for their sacrifice and to remain motivated in their mission, soldiers need to be recognised and appreciated by the society that they are serving. Shannon French explores the psychological need for soldiers to have the "profundity of their sacrifice" acknowledged. She explains that soldiers "are not mere tools; they are complex, sentient being with fears, loves, hopes, dreams, talents and ambitions" (French 2003: 10). If they are regarded as "mere means to an end", they risk losing faith in the society they are protecting and their performance on the battlefield will be adversely affected. This is particularly important because the soldier may face death at any moment, and he is therefore continuously reassessing the value of his life and potential sacrifice: it is the confidence of his own worth, anchored in the eyes of his community, which "gives life itself its value" (Coker 2007: 52).

Recognition therefore is crucial to the confidence and motivation of the soldiers: "they are public servants and derive much of their self-esteem from the extent to which they are esteemed by others, even civilians" (ibid.: 7). The judgement of society determines the soldier's sense of worth by giving value to his sacrifice and to the risks that he is taking. Honour and glory are important to the soldier as it is the hope of being welcomed home after the war that enables him to fight as his country requires: "it is important for them to conduct themselves in such a way that they will be honoured and esteemed by their communities, not reviled and rejected by them" (French 2003: 5). The soldier's craving for honour and glory has been manipulated by governments and military institutions and is used to drive these combatants to risk their lives on the battlefield. The celebration of soldiers and public rewards for acts of heroism and sacrifice provide a psychological impetus to those marching to war. Conversely, governments have also encouraged a culture of shaming 'cowardice' in citizens who fail to fulfil their civic duty, especially during a time of war. This was demonstrated in the UK during the First World War with the 'white feather' campaign which publicly branded any man who was not wearing a uniform.

Furthermore, killing is often antithetical to the instincts of the soldier who embodies "the values of the society in which they were raised and which they are prepared to die to protect" (ibid.). Ironically, clinical psychiatrist Jonathan Shay explains that "the painful paradox is that fighting for one's country can render one unfit to be its citizen" (Shay 1994: xx). Society's mandate to go to war therefore legitimises the murderous actions of the soldier and puts a moral stamp on killing. Coker argues that to be legitimate, killing has to be programmed, disciplined and directed and it must above all conform to the social construction of an enemy. For it is society which determines when a soldier kills, whom he kills, and even how he kills.

(Coker 2007: 61)

The approval of civil society therefore absolves the soldier of his actions. Finally, the military hierarchy eliminates any notion of moral responsibility, enabling the soldier to fulfil this bloody mission. The support of his family, community, and state is therefore necessary because they are the source of a soldier's moral salvation.

Soldiers not only crave society's recognition to fulfil their psychological and existential need: their physical survival also depends on the approval of their fellow combatants. Indeed, "the only thing that they fear is being shamed in front of their peers" (ibid.: 52). This is confirmed repeatedly in military literature by soldiers who explain that "an apprehension nagged (at each of us) that he might appear to be 'yellow' if he were afraid" (Sledge 1990: 5). Soldiers undergo extensive social conditioning that is "designed to keep soldiers at the front, not least of which is the belief that flight from danger will result in their comrades considering them a coward" (Moore 2009: 114). Any display of weakness or cowardice on the battlefield can endanger the entire group. Soldiers therefore share mutual responsibility and derive their only guarantee of security and potential survival by trusting that their fellow soldiers will equally act with courage and sacrifice while carrying out their own orders: "any man in combat who lacks comrades who will die for him, or for whom he is willing to die, is not a man at all. He is truly damned" explains William Manchester in his war memoirs. Coker explains this dynamic between soldiers who

put their lives on the line for each other (...). (They) have to trust each other to stand in line, rather than cut and run, to fight side by side and overcome personal fears, to go the extra distance. A soldier is expected to fight on even when all is lost rather than dishonour himself in front of his comrades.

(Coker 2007: 134)

This *esprit de corps* is realised through the rigorous training prior to battle, and through the constant presence of death during a war – the experience of which inculcates soldiers with "an ingrained sense of personal responsibility not to 'let the unit down'" (Moore 2009: 121).

Breaking the myth of combat: why soldiers fight

In approximately 12 BC, the Roman poet Horace wrote a poem in his *Odes* exhorting the young male citizens of Rome to join the military and fight for their country:

Dulce et decorum est pro patria mori:	What joy, for fatherland to die!
mors et fugacem persequitur virum	Death's darts e'en flying feet o'ertake,
nec parcit inbellis iuventae	Nor spare a recreant chivalry
poplitibus timidove tergo	A back that cowers, or loins that quake[1]

Horace's poem illustrates his society's celebration of patriotism and the vilification of cowardice. Ironically, Horace himself allegedly threw down his shield in panic and fled the battlefield at the Battle of Philippi in a bid to save himself from certain death. This is perfectly in line with Thomas Hobbes' explanation, however, that "men pursue their own self-preservation, which is why he agreed that a soldier might run away from battle provided he did so not out of treachery but fear" (Coker 2007: 62). Nonetheless, Horace's verse "*dulce et decorum est pro patria mori*" has been used repeatedly in government propaganda to glorify war and incite young men to enlist. Horace's actual behaviour on the battlefield, however, is rather representative of the stark realities on the ground. Wilfred Owen, a British soldier fighting in the trenches during the First World War, experienced the extreme horrors of that war and denounced the call for patriotism as "The old Lie" in a poem entitled 'Dulce et decorum est'.

Society expects its men (and sometimes, but rarely, its women) to go to war in its defence. A good citizen must have a sense of civic obligation or risk being stigmatised as "corrupt and vicious" (Locke cited in Coker 2007: 62). Propaganda, nationalistic sentiment and cultural mobilisation have contributed to a glorification of war among young men. Wilfred Owen berates society for telling this lie to its "children ardent for some desperate glory". The realities of war, which he and his contemporaries have described so vividly in poetry, literature and photography, are so horrific that it should in theory inspire soldiers to flee and not to fight.

The mobilisation of men and the courage of citizen-soldiers on the battlefields of Europe have been largely attributed to the patriotic zeal and moral superiority of the combatant. 'National soldiers' are considered to "have a personal stake in the matter" (Percy 2007: 122), which makes them loyal and effective warriors, ready to die for the cause. Experiences of soldiers on the ground, however, contradict societal expectations that its warriors are fighting out of patriotic fervour and suggest instead that it is the military infrastructure, with its draconian training, stringent laws and unbreakable *esprit de corps* that inspire the soldiers to fight on.

Lacklustre patriotism

Patriotism and civic duty are good for civilians supporting the war effort from their homes. They are also useful for inciting young men to join the army. In the battlefield, on the other hand, patriotism is a remote sentiment "rejected as fit only for civilians" (Graves 1929: 229). In his study entitled *The American Soldier*, sociologist Samuel Stouffer found that patriotism and nationalism were not major factors of motivation for soldiers fighting in the Second World War. In another survey, the social psychologist M. Brewster Smith concluded that less than 2 per cent of all American soldiers were fighting for patriotic reasons after they had landed in Europe during the war. The experience of war, with the constant presence of danger and potential death, forces the soldier to realign his values. Whereas he may enlist out of patriotic zeal, "there is no patriotism on the line. A boy up there 60 days in the line is in danger ever minute. He ain't fighting for patriotism" (ibid.). Brigadier Julian Thompson explains that fighting for an idea such as patriotism

> is a very fragile foundation on which to base morale, because in the stress of battle it evaporates. Whereas, if you're fighting for yourself, your comrades, for each other, that sustains you in the moments when you think you might be losing.
>
> (cited in Moore 2009: 246)

On the other hand, Leonard Wong argues that American soldiers fighting in Iraq have been motivated at least by *cause* if not by patriotism:

> once the war outcomes become apparent, the motivation shifts to more ideological themes (…). Despite the results of previous studies and the subsequent conventional wisdom that American soldiers are not motivated by ideological sentiments, many soldiers in this study reported being motivated by notions of freedom, liberation, and democracy.
>
> (Wong 2003)

Similarly, James McPherson explained that in the American Civil War, Confederate soldiers fought "for liberty and independence from what they regarded as tyrannical" (McPherson in Lindström 1994: 7).

Regardless of Wong's and McPherson's allegations, soldiers on the battlefield do not fight "for home, for the flag, for all the crap the politicians feed the public" (Hedges 2002: 38). The soldiers are not the public; they are the protagonists making the sacrifice that the public asks them to make. But the cost of war, particularly when the frontline is far from home, is borne by the soldiers who fear mutilation, death, capture and torture. And yet these men keep fighting, not out of patriotism, not just because they have to, but because their lives and the lives of the men next to them depend on it.

Esprit de Corps

The Air Force Officer's Guide states that "it is not primarily a cause which makes men loyal to each other, but the loyalty of men to each other which makes a cause" (cited in Janowitz 1960: 221). Small group loyalty or cohesion is purposefully developed in the armed forces because it is widely recognised that solidarity and group identity is the primary motivation for soldiers during combat: "by living together, sharing discomfort and danger and becoming utterly dependent on each other for survival, a unique bond of comradeship develops among these small bodies of men" (Moore 2009: 245). Armies, therefore, have repeatedly been organised into small administrative units, from the Roman Army's *conturbernia*, to Genghis Khan's ten man *arbans* and Frederick the Great's seven man *Kameradschaft*. Regiments are encouraged to develop a 'corporate identity' emphasising its uniqueness and lauding the previous heroic deeds of its adherents (Janowitz 1960).

The physical presence of comrades from the same background appears to be an important factor in keeping a soldier fighting and motivated on the battlefield. Australian Major Darren Moore explains that "when a soldier is beginning to succumb to such fears, often the realisation that others around him are still fighting provides enough motivation to fight on" (Moore 2009: 247). Wong also concludes from his study on American soldiers that "cohesion, or the emotional bonds between soldiers, appeared to be the primary factor in combat motivation" (Wong 2003). This is confirmed by Shannon French, who argues that the "sheer force of shared experience binds warriors together in the crucible of combat" (French 2003: 13).

Another important factor in this *esprit de corps* is its existential dimension that empowers the soldier. The nihilism of war can strip the combatant of his existential meaning, especially if he perceives that he is treated as a "mere means to an end" (ibid.: 10). Group mentality gives a soldier purpose. Soldiers must be made to "feel that although their individual contribution to the group may be small, it is still a critical part of unit success and therefore important" (Wong 2003). Furthermore, "a man's sense of his own worth is determined by the judgement passed on him by others" (Coker 2007: 52). Therefore the group's recognition of a soldier's instrumental purpose as an element of the unit is a self-affirming experience that validates the individual's presence on the battlefield. Coker describes the instrumental value of the soldier who, by risking his life

> may be brave but he must recognise also that he is more useful alive than dead. It is more useful for everyone if he is still alive to continue the fight. To lose one's life usefully is indeed to instrumentalise it.
>
> (Ibid.: 63)

Through constant exposure to one another and to high levels of risk, the soldier's immediate group becomes his family, sometimes even closer than his own core family. William Manchester, recalling his wartime experiences in the Pacific, explained that "those men on the line were my family, my home. They were closer to me than I can say, closer than any friends had been or ever would be" (Manchester cited in Hanson 2001: 119). The sacrifices and heroic deeds that soldiers perform on the battlefield are emotional and deeply personal acts of love directed towards a person who has come to mean more to the soldier than life itself: "devotion, simple and selfless, the sentiment of belonging to each other was the one decent thing we found in a conflict otherwise notable for its monstrosities" (Caputo 1996: xvii). The certainty of this devotion within the unit is what gives hope and existential meaning to the soldier on the battlefield: "that was part of what made it possible to do the job. You might be wounded, cut off from your comrades, surrounded by the enemy, but someone would be coming to get you" (Durant and Hartov 2003: 313).

This *esprit de corps* is a critical factor in combat performance. Group identity creates bonds of loyalty that guarantee as far as possible the continued presence of a soldier on the battlefield. The emotional support that soldiers receive from one another inspires an important amount of motivation, especially when the patriotic fervour and ideological aspirations have been blasted away by the harsh realities of combat. It is also an important factor in protecting combatants from psychological breakdowns by providing the structural and relational support that every individual needs, especially in times of hardship.

Fighting to go home

Charles Moskos argues that the emotional bonds tying soldiers to one another on the battlefield originates from a sentiment of self-preservation in a situation of heightened danger rather than from an altruistic concern for their comrades. Wong further explains that "once soldiers are convinced that their own personal safety will be assured by others, they feel empowered to do their job without worry" (Wong 2003). Soldiers on the battlefield have no choice but to 'bite the bullet' as the military and civilian infrastructure hold the threat of certain death if a soldier deserts his position.

In his research on combat motivation, Stouffer found that Iraqi war veterans claimed their motivation on the battlefield was stimulated by coercion, not out of solidarity or patriotic sentiment:

> their behavior was driven by fear of retribution and punishment by Baath Party or *Fedayeen Saddam* if they were found avoiding combat. Iraqi soldiers related stories of being jailed or beaten by Baath Party representatives if they were suspected of leaving their units. Several showed scars from previous desertion attempts.
>
> (Stouffer in Wong 2003)

Wong explains that this fear overrode camaraderie as the primary source of motivation because of poor training and low levels of group cohesion. His argument is substantiated by the threat that *all* soldiers face if they desert or surrender against the expressed wishes of their commander. Article 85 of the Punitive Articles of the Uniform Code of Military Justice states that

> (c) Any person found guilty of desertion or attempt to desert shall be punished, if the offense is committed in time of war, by death or such other punishment as a court-martial may direct, but if the desertion or attempt to desert occurs at any other time, by such punishment, other than death, as a court-martial may direct.

Threat of retribution and punishment, therefore, does not generally appear to be a significant motivation in keeping soldiers in the field. Officers have repeatedly used the threat of death on the spot to keep their troops from fleeing in panic during battles. A soldier therefore can choose between taking his chances on the battlefield, or assured death if he is caught deserting. Nonetheless, the high desertion rates on all sides during the Second World War suggest that this threat was not sufficient to keep men fighting if they chose not to: in the United States, 21,000 soldiers were convicted and sentenced for desertion; Nazi Germany executed 15,000 men during the war, but Joseph Stalin had 158,000 soldiers shot for deserting the Soviet war effort.

Whether soldiers fight out of a sentiment of self-preservation or out of solidarity, the outcome remains the same: only by surviving the battlefield will a soldier be able to go home. Their survival, whether physical, psychological, or moral, depends ultimately on the ties between the men fighting together on the battlefield. These emotional and practical bonds create a sense of loyalty between soldiers that keep them fighting on the battlefield even when all hope is gone. *Esprit de corps* or group cohesion is therefore a strategic objective that improves military performance and combat reliability.

Conclusion: the soldier's disenchantment

Until the end of the Cold War, security threats against the nation-state bore an existential dimension: wars were 'total', requiring absolute sacrifice from each and every citizen. Failure on the battlefield was perceived to have apocalyptic consequences: the experiences of entire populations at the hands of the Napoleonic, Nazi, Japanese, or Soviet armies brought up the cost of occupation and defeat. Citizens were consequently empowered with a common and mutual responsibility for the security of their state and the safeguard of their traditions. Governments exploited this opportunity by creating mass-conscripted armies, which fulfilled the dual purpose of bringing the population under the state's control, and improving the security of the nation by affordably mobilising all necessary resources and manpower.

The security environment changed in later decades of the twentieth century, however. In the first place, the fall of the USSR in 1989 reduced the existential threat of a nuclear attack. The threat of invasion has also been reduced significantly, with a greater institutional respect for territorial sovereignty which is safeguarded by supranational organisations such as the European Union and the United Nations. Security threats have shifted away from an existential and national dimension and now encompass more value-minded missions such as peace and humanitarian operations. Furthermore, there has been a decline in inter-state wars with a corresponding rise in civil wars and state collapse. The military purpose of armies has shifted from "war fighting or war deterrence to military deployments for peace and humanitarian purposes" (Moskos *et al.* 2000: 3). In this security context, the existential rationale for maintaining huge armies of citizen-soldiers has disappeared.

Not only has the rationale for huge armies disappeared, "the non-existential nature of most contemporary security problems decreases citizens' willingness to accept large military budgets and to contribute personally to national defence" (Krahman 2010: 245). The armed forces have also lost their 'glamorous' image. Even in the United States, which appears to many as the most militaristic and patriotic nation today, "there is a finite number of competent people willing to choose a career that requires wearing a uniform, performing often dull work, such as guard duty, with alertness, and being ready at any moment to risk one's life for others" (Moskos and Glastris 2001). In *The Warrior Ethos*, Coker explains that society no longer celebrates war and its warriors: it "can no longer interpret sacrifice, except as a waste of life" (Coker 2007: 102). Parents are reluctant to allow their children to go to war – let alone encouraging them to do so; and society carries on 'as usual' while battles wage on faraway frontiers. For the soldier, whose sanity and performance depend on an expectation of recognition, this utter lack of support and gratitude is a killer blow.

Faced with the non-existential security environment of the twenty-first century, Western countries have abandoned the draft in favour of smaller and more professional armies: the United States abolished the draft in 1975, whereas France suspended conscription in 2001 and Germany in 2011. Only six Western countries still operate a system of military service. On the other hand, their most likely military and economic competitors have maintained conscription as a national policy: Egypt, Brazil, China, Algeria, Russia, Iran, Venezuela, Israel and North Korea continue to enlist their young men (and women) into the armed forces. In the West, as wars have lost their existential and patriotic glory, armies are rapidly feeling the pinch of a shortage of manpower.

The civic-militarism that enabled the massive armies of the twentieth century is presently undermined by the murderous legacies of French, German and Japanese nationalism and ethnic militarism in the Balkans (among others). Civic duty is also eroded by the effects of globalisation: improvements in communications and transportation technology along with the high rate of mobility and migration have facilitated a system of overlapping identities and loyalties that are no longer centred on the nation-state (Bull 1977). Furthermore, supranational and regional organisations have weakened and diminished the influence of the central state. These events have significantly affected the social contract, replacing social and military responsibility with absolute and unconditional rights ranging from health care to security. The apathetic attitude of citizens towards their nation is evident not only in their refusal to attend to their military duties, but also in the diminishing rate of voter turnout: only 10 countries today enforce compulsory voting. In other countries, voter turnout is as low as 39.79 per cent in the 2007 Swiss Parliamentary election and 57.47 per cent in the 2008 US Presidential elections. In France, the United Kingdom and Germany, less than 75 per cent of the population bother to vote in general elections.

Max Weber argued that industrialisation and the rationalisation of modern life led to a "disenchantment of life" which he defined in terms of the erosion of sublime values from public life (Weber 2004). The citizen-soldier has every reason to be disenchanted with war. There is no longer any patriotic fervour to encourage citizens to enlist. Military budgets are increasingly being cut, forcing armies to take shortcuts and leaving soldiers under-equipped in the heat of battle. The soldier's sacrifice is unwarranted, undeserved and unrecognised. There is no existential threat or existential purpose to justify the soldier's presence on the battlefield. Soldiers have lost track of the mysticism of war. Unlike mercenaries who generally fight for money or for the love of war, there is very little incentive for soldiers to keep fighting. On the other hand, there is a crucial distinction between these two combatants. No matter how disillusioned they may feel, only the national soldier can be counted on to fight to the very end: the punitive juridical system of the military and the *esprit the corps* deliberately engineered between the soldiers make these men brave beyond reason – a quality that can by no means be expected from mercenaries.

Note

1 Conington, John. "The Odes and Carmen Saeculare of Horace by Horace". *Project Gutenberg*. N.p., Apr. 2004.

4   The French Foreign Legion

The legionnaire is the mercenary-soldier *par excellence*. This modern-day Achilles fights to the death for a cause that is not his own, while operating under a strict military code and being entirely answerable to the French state. Born out of the ideological flames of the French Revolution and thrust into the era of the citizen-soldier, the French Foreign Legion has survived and thrived against all odds. Indeed, the Legion is unique in its heterogeneity and in its success at overcoming institutional and societal norms against mercenaries and foreigners. Of its 7,650 legionnaires, 75 per cent are foreigners representing 140 nationalities integrated into the *Armée de Terre* of which it makes up 5 per cent of the manpower. The *Képis Blancs* are recognisable worldwide and are part of their host country's military self-image. Furthermore, today's legionnaires embody and inspire the modern-day warrior in a world that appears increasingly disenchanted with war.

This chapter traces the history of the French Foreign Legion and explores the identity and motivations of the men who join as legionnaires. The study aims to inform the discourse on the hybridisation of the armed forces by highlighting the ways in which France has, in this case, managed to overcome societal norms and military expectations regarding nationality and patriotism. Through a stringent use of disciplinary methods, an ideology focused on solidarity and duty and the forging of a new identity anchored in its own traditions and history, the Foreign Legion has managed to build a regiment whose performance on the battlefields have inspired the respect and admiration of their peers. Consequently, the French Foreign Legion has become '*le plus beau corps de France*', successfully integrating foreigners and French citizens into an army whose national pride has been a benchmark for citizen-armies since the nineteenth century. Lessons from this case study can be brought forward to address some of the problems facing modern armies that seek to merge nationals and foreigners or soldiers and private contractors in their military expeditions.

A military experiment

In 1830, France was a haven for political refugees and unsuccessful revolutionaries. The ideals of the French Revolution attracted political activists from Russia, Spain, Italy, Poland and Belgium who found asylum in the country of freedom and equality. A second Revolution in July 1830 saw the Bourbon monarchy over-thrown and France unilaterally renouncing all extradition treaties. Consequently, foreign immigrants flocked into the country for political or economic reasons and, finding themselves unemployed, "became vagabonds, delinquents or even con-tributed to the considerable political turmoil of the period" (Porch 2010: 2).

Threatened by the presence of these turbulent personalities on French ter-ritory, the newly crowned King Louis-Philippe sought to control these foreigners by integrating them into the armed forces. Unfortunately for the king, a fervent anti-mercenary norm had developed out of the two revolutions: Article 13 of the 1830 Charter stated that "*aucune troupe étrangère ne pourra être admise au service de l'État qu'en vertu d'une loi*" (no foreign troop will be admitted into the service of the State unless exempted by decree). This stipulation was intended as a safeguard against the king's inclination to protect himself from the French people by using foreign troops, generally the Swiss Guard. It was by royal decree therefore that the Foreign Legion was established on 9 March 1831 with the recommendation that they should only serve outside of France.

Despite the Legion's claim that the 1831 ordinance follows France's long tra-dition of using foreign troops, the circumstances surrounding the birth of the Foreign Legion indicate otherwise: France's experience with foreign troops had made these combatants terribly unfashionable. In 1832, a new recruitment bill, the Loi Soult, declared that "*Nul ne sera admis à servir dans les troupes française, s'il n'est Français*" (no one shall be admitted into the French troops if he is not French), emphasising the general distrust vis-à-vis arming foreigners. French War Minister Maréchal Nicolas Soult wrote in 1834 that

> as the Foreign Legion was formed with the only purpose of creating an outlet and giving a destination to foreigners who flood France and who cause trouble (...) the government has no desire to look for recruits for this Legion. This corps is simply an asylum for misfortune.
>
> (Azan 1936: 124)

Donald Porch describes the birth of the Legion as

> the illegitimate child of the July Revolution, an embarrassment, at once acknowledged and shunned, whose meagre patrimony was to be the right to die for France in the wastes of her empire. Its very name suggested all that was unfamiliar, unknown and distrusted, the very antithesis of the citizen, the compatriot, the comprehensible.
>
> (Porch 2010: 6)

Early days

With such an inglorious beginning, the Legion's early years were inevitably fraught with difficulties. The men who made up the legion were a heterogeneous group. Captain Morin de la Haye describes the legionnaire as a *déclassé*, an adventurer, someone made bitter by life, a bandit. In the Legion we get everything. In the companies which have a strength of three to four hundred men, one finds a mixture of ex-officers, ruined gentry, anarchists and freed convicts

(Richard 1890)

Finding competent officers to train and lead the legionnaires was a task in itself. French officers were reluctant to take command positions in the Legion and nomination to such an assignment was considered as punishment. The first commander of the Legion, Swiss Baron Christopher Antoine Jacques Stoffel, himself criticised for lacking command experience, complained in June 1831 that the Legion had only eight competent officers out of 26. Non-Commissioned Officers also frequently lacked training, confidence and authority, undermining the organisation of the Legion and encouraging a tendency towards insubordination and indiscipline.

French commanders initially dismissed the Legion as a combat unit and assigned the battalions to the least appealing and most futile tasks, including manual labour and road-building duties. The legionnaires were generally ignored by the French Army who regularly forgot to send them reinforcements, pay and even food. Between 1831 and 1835, the Legion lost about a quarter of its numbers to disease. In its early days in Algeria, the legionnaires spent more of their time building roads or in convalescence than in active combat.

The Legion's perceived expendability was most evident in the Spanish campaign. In 1833, Spain was torn apart in a civil war between the supporters of the three-year-old Queen Isabella and her reactionary uncle, the Infante Carlos, who refused to recognise a female sovereign. Isabella's reign was supported by the Quadruple Alliance made up of the United Kingdom, France, Spain and Portugal. As the Carlists grew stronger, the Regent Maria Christina appealed in her daughter's name for military support. France's difficult experiences in Spain during the Napoleonic Wars had left the government hesitant with regards to committing itself to another unpopular campaign in the perilous Iberian Peninsula. Therefore on 28 June 1835, King Louis-Philippe compromised by ceding the French Foreign Legion to Madrid. In this manner, France provided diplomatic support to Isabella short of sending French troops. Furthermore, by leasing the Legion to the Spanish authorities, the French government declined any responsibility towards their troops, and highlighted the expendable nature of this undesirable army.

The leasing of the Foreign Legion was badly received by the soldiers who complained that they had come to France to serve France, and not other countries. The Legion was not so much 'loaned' as auxiliaries, which had been the practice in armies throughout history, but given over entirely to the Spanish. It was even temporarily renamed "the Spanish Legion". The French government argued that, as mercenaries, it should not matter to the legionnaires who they were fighting for or where their salary came from – although Spain was expected to pay the soldiers, something which they often forgot to do. The Legion was effectively treated as a force for hire, owned by France and given to Spain with diplomatic relations serving as the *monnaie d'échange*. In contrast, the United Kingdom sent to Spain a force of 4,000 to 10,000 volunteers from the British Army line regiments between 1835 and 1837. The British troops were known as the Auxiliary Legion, and as volunteers, also released Britain from any moral or material commitments while demonstrating its diplomatic support to Isabella's cause. Both the British Auxiliary Legion and the French (now Spanish) Legion suffered heavy losses in the Carlist wars. By 1837, only 1,500 British soldiers remained in Spain, with some 2,500 casualties – a quarter of the force. For the French legionnaires, the war was much worse: out of 5,000 legionnaires dispatched to Spain, only 159 returned to France in 1839. The Foreign Legion had effectively been sacrificed – politically and militarily – to Spain.

Earning its laurels

When the French authorities abandoned the Foreign Legion to the savagery of the Spanish War, the military experiment of a regiment of foreigners appeared to come to its (un)natural end. The colonisation of Algeria in 1830, however, along with the continued flow of immigrants into France, offered a new wind to the Foreign Legion. On 16 December 1835, King Louis-Philippe created the *nouvelle Légion* with the purpose of supporting France's colonial project.

By 1840, the Legion had enough recruits to form two regiments. The Algerian campaign was harsh, requiring exceptional stamina from the legionnaires. The dynamism of the postings in Algeria transformed the Legion from an army of labourers to an army of combatants. The opportunity for military engagements restored the morale of the troops and contributed to their improved performance, born out of discipline and hardship. A handful of exceptional leaders, most notably the governor-general of Algiers, Thomas '*le père*' Bugeaud, managed to develop a system of command and control that sowed the seeds of a regimental spirit.

Algeria eventually became the Legion's base and is engraved in the history and identity of the army. The legionnaires built their military headquarters themselves in Sidi-bel-Abbès, a town that would be the home of the Legion until 1962. The "glorious" reputation of the legionnaires was forged in the French colony, with major battles including Constantine (1837), Djidjelli (1839), Millianah (1840), Zaatcha (1849) and Ischeriden (1857). The successes of the Legion in Algeria led to its status as a colonial army, with exceptional endurance and mobility.

The Legion was more than just a colonial army, however. It was called to fight besides regular French forces in the Crimean War from 1854 to 1856. The Crimean campaign was particularly important for the Legion, because it gave it a "double vocation; that of fighting in France's European wars as well as in her colonial wars" (Porch 2010: 128). In this way, the Legion became a vital component of France's system of defence. The 1859 Italian campaign confirmed France's acceptance of the Foreign Legion as a fighting force: in the battle of

Magenta, the *2e étranger* forged the reputation of the troops by fighting with legendary courage. The Legion subsequently accumulated battle honours in the 1850s, reaching new heights in the colonial wars of the 1880s as the French government tripled the Legion's recruitment numbers and transformed the force into the striking arm of the colonial army.

The colonial project

In 1883, the Foreign Legion landed in Tonkin to support the French invasion of Indochina. The Legion was part of the attack column that captured the cities of Son Tay and Tuyên Quang and was also an integral component of France's victory in the Sino-French War. By March 1885, the French had secured large parts of the country despite having relatively few troops at hand. Indochina was to become a 'second fatherland' for the legionnaires: in Tonkin, there were opportunities for action. By contrast, Algeria had been pacified and there were few excuses for military engagements. Furthermore, pension pays were doubled in Indochina where legionnaires qualified for 'colonial pay', a privilege denied in Algeria as it was considered to be part of metropolitan France. Indochina was perceived as an exciting posting, with guaranteed battles, an abundance of women, and servants to take care of menial tasks.

The French colonial project continued and saw the Foreign Legion sent out to conquer parts of Africa, namely Dahomey (1892), Western Sudan (1892), Madagascar (1895) and Morocco (1900). The Legion's ability to adapt to any terrain was proven in its battles from the jungles of Indochina to the deserts of Morocco and Algeria. The Foreign Legion became the *outil de base* for France's colonial ambitions, even more so when, following the Franco-Prussian War, French public opinion turned against imperial expansion: colonialism became perceived as a waste of French lives and a distraction from the German threat. France compensated by using native troops in her colonial expeditions, but the commanders requested the participation of the Legion because they were perceived to make better combatants. Furthermore, the French did not trust their native troops and preferred to deal with white soldiers. The legionnaires were therefore employed to discipline the troops as much as to fight the enemy.

The Foreign Legion was to be the beacon of French colonialism until the process of decolonisation in Indochina and Algeria from the 1940s to the 1950s squeezed the Legion out of its military bases. The Legion's last years in Indochina were characterised by a steady deterioration of the French imperial troops, a shortage of manpower, incompetent leadership and gross mistakes, which eventually culminated in the army's final defeat at Dien Bien Phu in May 1954. The difficulties on the ground were exacerbated by the apathy of the French towards another war so soon after its own struggles in Europe. The long war of attrition in Indochina cost the Legion 11,000 casualties, which corresponded to about one-third of its manpower in the country. Nonetheless, Dien Bien Phu is celebrated by the Legion as a paragon of courage and sacrifice and a display of heroic resistance.

The Legion's return to Algeria was a sombre event. Humiliated, defeated and abandoned, the legionnaires began to lose morale. This was exacerbated by the French policy of deducting the food allowance of former prisoners of war from their back pay on the assumption that the legionnaires "had been fed by the Viet Minh during captivity and therefore were not entitled to it" (Porch 2010: 566). The legionnaires were increasingly isolated from their political base, particularly as France set Tunisia and Morocco on the path to independence, paving the way for Algeria, a move that the Legion refused to accept.

The loss of Algeria and the Legion's rebellion

The outbreak of the Algerian War in November 1954 caught the French government by surprise. Adjudant-Chef Janos Kemencei expressed his anger that the equipment provided by the authorities in 1955 to fight against the National Liberation Front (FLN) was utterly unsuited for military operations and appeared to be leftovers from the First World War. In his memoirs, he wrote

> I no longer had any faith. My professionalism remained intact. But I no longer wanted to fight for causes lost in advance, like here in Algeria (...). To support an intemperate climate without appropriate material, support the constant absence of hygiene, swallow on operations execrable flood, all of this had literally disgusted me with the army.
>
> (Kemencei 2000: 318)

The French Army, including parts of the Legion, ran counterinsurgency operations in Algeria, characterised by a brutal war of attrition with escalating levels of violence, particularly between the Legion's *paras* and the local FLN terrorists. The unrestrained use of torture, however, led to public condemnation in France, particularly among the literary community.[1]

The Legion was particularly provoked when, despite promises to safeguard Algeria from independence, President Charles de Gaulle reached out to the FLN. Ignoring his own declarations of support for an *Algérie française*, in 1960, de Gaulle yielded to the demands of the United Nations and began secret peace talks setting out the conditions for Algerian independence. This provoked the Legion, particularly in the wake of its recent experiences of defeat and abandonment in Indochina. The loss of Algeria, although controversial in France, was unacceptable to the Legion whose *ville sainte* was still Sidi-bel-Abbès, which the Legion had built itself. Porch explains that "as it increasingly appeared that Algeria would be cut free from France, unit loyalty and solidarity became an important factor in determining attitudes, foremost among them the feeling that without Algeria there could be no Legion" (Porch 2010: 607). Morale was severely undermined as the legionnaires feared for their future. The legionnaires of the 1er REP were particularly sensitive to the loss of Algeria. They had been almost constantly in battle since 1940 and had suffered from huge losses in Indochina, which they attributed to the betrayal of the French government. The regiment's headquarters in Zéralda also placed them in the political heart of Algeria.

On 22 April 1961, the 1er REP under the leadership of General Maurice Challe seized control of Algiers in an attempted putsch against President de Gaulle. The French president responded by calling a state of emergency and demanding the arrests of select officers. The coup was not supported by the French in Algeria or in France, and led to a mobilisation of the unions, conscript soldiers and intellectuals. The coup also divided the Legion, with Lieutenant Colonel Brothier declaring that "the putsch is a French affair; it is unthinkable that foreigners should become mixed up with it" (Le Mire 1982). His statement was to save the Legion from being abolished completely. In his own account, 1er REP British Legionnaire Simon Murray observed that "the army is completely divided and we appear to be very much a minority ... I wonder what will happen if this *putsch* does not succeed and what our own position will be" (Murray 1978). On 26 April, French police and military units surrounded the 1er REP and arrested General Challe. The 1er REP was dissolved on 30 April upon the orders of Defence Minister Pierre Messmer. The legionnaires tore down the barracks that "they had built themselves from scratch, brick by brick, in the tradition of the Legion" (ibid.: 139) and left their base in Zéralda singing Edith Piaf's song "*Je ne regrette rien*".

Although the failed coup had been engineered by French officers, the Legion was made the scapegoat of the whole affair. The formal independence of Algeria on 3 July 1962 forced the Legion out of the base that had been its home for 120 years and left the French Army temporarily without a role in Africa. Simon Murray described the problem for the Legion: "suddenly there is no purpose, there is no direction. Bewilderment is quickly superseded by boredom, which is itself overtaken by a rapid decline in morale. Discontent follows and the system begins to rot" (Murray 1978: 190). As a result of this episode, the Legion's numbers was cut from 40,000 soldiers to 8,000. Headquarters were moved to Aubagne in the South of France and the Legion began its search for a new meaning and role in a post-colonial world.

The transformation of the Legion

On 10 December 1962, the fate of the Legion was suddenly decided. After rumours of disbandment and months of building roads, the French Inspector-General of the Legion, General Jacques Lefort, announced to the 2e REP that they were to be transformed into a rapid reaction force. The geopolitical situation had changed with the end of the colonial era, and France subsequently had fewer requirements for troops as a whole. In order to survive in this new competitive environment, the Legion had to clean up its image and diversify its skills. Hence the regiments were required to "set about training up specialised sections in underwater combat, demolition, guerrilla warfare, night fighting, special armaments" (ibid.: 211). Simon Murray enthusiastically described the training programme ahead:

we will be trained to operate tanks and armoured vehicles, we will be taught to ski and mountaineer, we will become familiar with submarines, we will be sent on survival courses and we will become highly skilled and dextrous with multipurpose capability. This is all terrific stuff (...). We're back in business. Somebody thinks we can do more than just build bloody roads all day.

In 1969, the Legion received its first post-colonial posting in Chad when the French government sent them to support President Tombalbaye's regime against a rebel insurgency. This was followed in 1976 by a second mission in Djibouti where the 2e REP was sent to release a bus-load of children that Somali terrorists had taken hostage. The new role of the Legion, however, was cemented in Kolwezi in 1978. A mining town in Zaire's mineral-rich province of Shaba, Kolwezi was overrun by 4,000 rebel fighters who crossed the border from Angola and took hostage the 3,000 European engineers and families living in the town. President Mobutu appealed to the French and Belgians for help, and on 19 May 1978, 650 *paras* were dropped over Kolwezi with orders to push back the rebels and release the European hostages. The mission was lauded as "a resounding success for 2e REP, a vindication for the Legion's new role" (Gilbert 2010: 243). However, five legionnaires were killed and 25 were wounded, with an additional 190 Europeans and 200 black civilians killed by the rebels as they retreated back to Angola.

The weak military and political capabilities of the new post-colonial African states provided a fresh opportunity for the Legion: France sought to maintain its influence in the former francophone colonies and, upon invitation, sent its legionnaires to help restore and maintain order in disputed territories. Consequently the Legion continued to see action in Chad, where it returned in 1978 and remained until 1988, and in Djibouti, formerly known as French Somaliland; it helped evacuate French civilians in Rwanda, Gabon, and Zaire in 1991, returning to Rwanda with *Opération Turquoise* in 1994. France sent its Legion to Centrafrique in 1996 and to Congo-Brazzaville in 1997 to evacuate the French expatriate community following violence in the area. A coup attempt and rebellion in Côte d'Ivoire in 2002 brought the French back into the country as President Laurent Gbagbo asked France to implement a ceasefire. The 2e REP, 2e REI and 1er REC were subsequently sent to Côte d'Ivoire to support the French troops in *Opération Licorne*.

French military activity in francophone Africa has become a cornerstone of its foreign policy, known as *Françafrique*. This term encompasses the alleged neo-colonial interests of France in its former colonies, and is played out through diplomatic relations, economic aid and investments, and military intervention. The Foreign Legion has been France's military arm in Africa, and therefore played a central part in the country's continued international prestige. French aspirations to remain a great power in the world have been played out in Africa: as Foreign Minister Louis de Guiringaud once said, "after all, Africa was the only continent where France could still change the course of history with a few hundred men" (cited in Taylor 2010: 66). Indeed, former President François Mitterand ominously declared in 1957 that "without Africa, France will have no history in the twenty-first century" (ibid.).

The Foreign Legion has not been limited to Africa, however. As part of its new role and in view of France's increasing collaboration with the international community, the French Foreign Legion has been called upon to participate in peacekeeping missions under the banner of the United Nations. The 1er REC was included in France's military contribution to the Allied invasion of Iraq in 1991. In 1992, the Legion was sent to Cambodia and Somalia to facilitate humanitarian efforts. This was followed by the controversial *Opération Turquoise* in Rwanda, where French forces, working apart from the UN, were accused of facilitating genocide by providing a safe haven for Hutu *génocidaires*.

The Legion's new role as a contributor to international military cooperation was not easily accepted by the legionnaires. The civil war between the Serbs, Croats and Bosnians in ex-Yugoslavia brought in the 6e REG and 2e REP, but the restrictive conditions of the UN mandate frustrated the legionnaires, who felt that their potential was under-exploited. Legionnaire Matt Rake described the difficulties that the Legion faced while working with the UN:

> this was the first U.N. mission the Legion had ever done. We had to change our berets to the blue U.N. ones. This didn't go down well – we'd signed up for the Legion, and not the U.N. And we were made to wear the French flag on our arm, which we'd never done before, and that didn't go down well either.

> (Rake cited in Gilbert 2010: 247)

The operational restrictions imposed on the Legion further undermined the *raison d'être* and traditional autonomy of the legionnaires. Pádraig O'Keeffe expressed his personal frustration with the mission in Bosnia:

> our mission in Bosnia was a joke – as far as I was concerned we were helping no one. We knew it, and so did the Serbs, the Bosnians and the Croats. We were glorified aid workers. We had been trained to take military action, we knew the action that was required, and yet we were told to sit on our hands. And while we stood by, innocent people died. I was sick of all the bullshit.

> (O'Keefe 2007: 121)

Nonetheless, the Legion's role in the twenty-first century is anchored in its ability to contribute to international peacekeeping, humanitarian and disaster alleviation efforts. In 2002, France joined NATO forces in Operation Enduring Freedom and committed its legionnaires to working as minesweepers and as instructors for the new Afghan Army. In 2008, following a new outbreak of violence in the area, the Legion was sent back to Chad with EUFOR, a multi-national EU force. The Legion's mission in Chad was to facilitate the delivery of humanitarian aid and protect civilians and UN personnel. The Legion also participated in international disaster alleviation efforts in South and South-East Asia after the 2004 tsunami in the Indian Ocean. Currently, legionnaires are working as peacekeepers in Lebanon with the United Nations Interim Force in Lebanon (UNIFIL).

The French Foreign Legion has changed dramatically since its creation in 1831. From a small force whose purpose was to channel 'undesirables' out of France to the country's most elite and celebrated regiments with an international reach, the Legion has gloriously weathered public opposition, scandals, desertions, the end of the colonial era, and the perpetual transformation of French foreign policy. With its ability to adapt, its excellent combat record, its non-military qualities and multi-lingual capabilities, the Legion remains an asset to France's military operations within a new framework of global cooperation.

The legionnaire

The legionnaire is a soldier who has voluntarily enlisted in the French Foreign Legion, and is therefore part of France's *Armée de Terre*. Legionnaires must be male, between the ages of 17 and 40, and are required to meet the Legion's physical requirements. In practice, however, the Legion has historically been a channel to clean France of undesirable foreigners and sometimes even undesirable Frenchmen. In 1834, an inspector for the Legion complained that "in their haste to speed undesirable foreigners out of France, mayors and recruiting officers were enlisting men who were visibly infirm" (Porch 2010: 14). Men accused of petty crimes in France were also sometimes given the choice between going to jail and joining the Legion. Originally, legionnaires were either foreigners who were fighting for a country that was not their own, or renegade Frenchmen.

Today, the Legion is an elite unit of the French Army, but 75 per cent of its manpower is foreign. This belies the general expectation that conscripts and national forces are superior combatants compared to foreigners or mercenaries. Despite its reputation as an exceptional combat force, however, the Legion suffers from above average desertion rates, a characteristic typical of mercenary units, as described in the previous chapter. Nonetheless, the legionnaires have defied conventions and proved their mettle on the battlefields of French history, earning the admiration and respect of the French public and its armed forces.

Motivation

Political repression and poverty have generally been the main motivations for men enlisting in the Foreign Legion. Erwin Rosen, a legionnaire in 1905, was shocked by the physical state of the men waiting at the recruiting office in Belfort: "a dozen men were there. Some of them were mere boys, with only a shadow of beard on their faces; youths with deep-set hungry eyes and deep lines around their mounts; men with hard, wrinkled features telling the old story of drink very plainly" (Rosen 2010). Another legionnaire, Adrian Liddell Hart, explained in 1951 that for a large proportion of men, the Legion was "a sanctuary in the most literal sense. A few of them, even today, are escaping from their police for civil offences. Many more are escaping from their governments for political reasons" (Hart 1953: 203). It is therefore not surprising to find that the nationalities of the legionnaires follow the political upheavals of history: White Russians enlisted in large numbers in the 1920s, Germans volunteered *en masse* after both World Wars, Hungarians and other Eastern Europeans joined the Legion with each new Soviet invasion. Military coups and revolutions filled the recruitment bureau of the Legion. Colonel Robert Devouges said in 1981 that the Legion "reflects the troubles of the world. Laos, Cambodia, Bangladesh – you name the event, and we'll have the men" (Gilbert 2010: 19).

The Legion's policy of anonymity has been essential for men fleeing a political or criminal past. The *identité declarée* was originally a legally endorsed recruiting mechanism that facilitated immediate enlistment in times of conflict, as foreigners were not required to prove their identity. Men signing up for the Legion used a *nom de guerre* and could take on any nationality of their choice. This was particularly useful for Frenchmen who, until 2007, were disallowed from enlisting, except as officers: "on paper, apart from the officers, there are no French nationals serving in the ranks. Any Frenchmen are listed as Belgians, Swiss or French-Canadians" (Salazar 2005: 14). The anonymity of the Legion, however, was also an opportunity for men who had suffered difficulties and failures to gain a second chance in life. The Legion vehemently protected the identity of its legionnaires during the entirety of their service. The *identité declarée* became a tool of social inclusion and military integration, and forged a silent debt between the Legion and its men: "*leur fidélité, leur dévouement, leur engagement et leur disponibilité sont à l'image de cette volonté radicale de rupture et de don exclusif à leur famille d'accueil*" (their fidelity, their devotion, their engagement and their availability are at the image of this radical desire for rupture and this exclusive gift towards their new family). It was a symbol of hope and new beginnings, but also a statement of dependency and loyalty to the Legion who had generously endowed its soldier with this priceless gift. Since 20 September 2010, legionnaires are no longer required to enlist under a *nom the guerre*, although it remains an option and they may at any time reclaim their true identity.

Money has never been a source of motivation for legionnaires: "during the nineteenth century – and well into the twentieth – pay was so poor that it was only the most desperate who volunteered for economic reasons: in order to put clothes on their backs and food in their mess tins" (Gilbert 2010: 16). In fact, legionnaires were paid about five centimes per day, so little that it "bought nothing more than a box of matches, and which even the Arabs scorned" (Porch 2010: 188). The French government argued that "to pay legionnaires a fair wage would open them to charges of being mercenaries. Therefore, harsh administrative logic decreed that the virtue of the legionnaire, as well as that of France, was somehow redeemed by his impoverishment" (ibid.: xiv). The pay was so low that legionnaires were at times forced to sell their equipment to pay for food and medication. This changed, however, in the 1960s, when the salaries of the Legion were improved to increase its appeal to a wider selection of economic migrants. Today, a new recruit can expect a monthly salary between €1,043 and €3,567 depending on his posting.[2] This includes housing, food and equipment. All legionnaires enlist as bachelors, regardless of their marital status or whether they have dependents, and therefore the only variation in salaries is according to experience and postings. Only after five years of service may legionnaires seek permission to marry.

The Foreign Legion has traditionally appealed to professional soldiers and men with a military vocation. Adrian Liddell Hart explains that many recruits are "men who want to be professional soldiers and cannot soldier in their own countries, or have decided that it is anyhow better to soldier in the Legion" (Hart 1953: 203). The Legion is a sanctuary and an opportunity for new beginnings; it offered its soldiers a community that is recognised by the entire world, claiming a mystique that went beyond a normal military force. It is therefore a self-glorified rite of passage in which its legionnaires are promised an opportunity to test their manhood in combat.

The Foreign Legion boasts several prestigious legionnaires and officers in its history: King Peter of Serbia and French Prime Minister Pierre Messmer were both officers of the Legion. It counts among its roster of legionnaires Shapour Bakhtiar, the Shah of Iran's last prime minister; Prince Aage of Denmark; Prince Amilakhvari of Georgia; British multi-millionaire businessman Simon Murray; and French nationalist leader Jean-Marie le Pen. The Legion has traditionally attracted romantics, artists, writers in search of existential experiences and emotional challenges, although "the idealistic adventurer (...) was actively discouraged" (Gilbert 2010: 14). German painter Hans Hartung, German writer Ernst Jünger and the Hungarian intellectual Arthur Koestler are to be included in the Legion's hall of fame. The Legion has been the home of men from different cultures, countries, races and religions. The diversity of its soldiers has given the Legion a unique personality. British legionnaire A. R. Cooper describes the Legion as

> a refuge, a meal ticket, a place for rehabilitation. It can also be a profession. A man goes to it without identity papers, with the nationality of his choice, and shorn of criminal records. He leaves his past outside the recruiting office door.
>
> (Cooper 1969: 74)

According to Porch, the Legion is "an honourable asylum, (offering) the possibility of satisfying either their tastes or their needs, and finally, while serving the cause of civilisation, the promise of a modest future with a pension" (2010: 85). Men join the Legion for many reasons, although clearly they can rarely be motivated by salary or wealth. Nonetheless, as foreigners, legionnaires have been viewed with suspicion and are often considered little better than mercenaries who fight for a country that is not their own, with no loyalty or patriotism to ensure their performance on the battlefield.

Death and the legionnaire

The Legion has historically been used as cannon fodder by the French: sent into the most dangerous places, sacrificed to the enemy and neglected by France. This was symbolically voiced by General Oscar de Négrier who told the legionnaires leaving to go to Tonkin in 1883: "*Vous, légionnaires, vous êtes soldats pour mourir, et je vous envoie où l'on meurt!*" (Wellard 1974: 11) (You, legionnaires, you are soldiers to die, and I am sending you where one dies). It is no surprise, therefore, that legionnaires have adopted a nihilistic attitude towards their own life and death, evidenced in the high suicide rates and the practically suicidal missions in which they take part. In the Legion, "the majority of its members consider themselves already dead. That is the full evolution of the soldier (...). Dying is what the Legion is all about" (William Brooks cited in Porch 2010: 625). The *sappeurs* of the French Legion embody this philosophy: they are tasked with engineering duties including the destruction of the enemy's heavy machinery while under fire. They traditionally had a very short life expectancy and consequently were given more leeway in the army – which is why they symbolically wear a beard as mark of their special privileges and as evidence of the high risks they take and the sacrifice that they make.

Embracing death is part of the code of the legionnaire. Article 6 of the Honour Code states that "*La mission est sacrée, tu l'exécutes jusqu'au bout et si besoin, en opérations, au péril de ta vie*" (Your mission is sacred. It is carried out until the end, in respect of the law, the customs of war, International Conventions, if needs be, at the risk of your own life). Upon entering the Legion, the recruit swears an oath to defend himself and his comrades to the death, if necessary. The Battle of Camerone on 30 April 1863 exemplifies this philosophy: the Foreign Legion was stationed in Mexico in 1863 as part of France's campaign against President Juárez who had suspended interest payments to its creditors, France, Britain and Spain. During the campaign, the French sent a convoy carrying three million francs, equipment and munitions. A reconnaissance team of 62 legionnaires and three officers were sent out ahead of the convoy. The men were attacked by up to 2,000 soldiers from the Mexican cavalry and infantry. Led by Captain Danjou, the legionnaires refused to surrender, despite the inevitability of their impending defeat, earning the admiration and respect of their opponent. After 11 hours and having lost all but three men, the Mexicans allegedly begged the legionnaires to surrender and save themselves, which they did on conditions that their fallen comrades would be collected and that they would be allowed to keep their weapons. The heroic sacrifice of Danjou and his men saved the convoy, and set an example for the legionnaires. Camerone Day on 30 April is one of two sacred holidays in the Legion's calendar. A huge parade is organised to showcase the Legion and the famous wooden hand of Captain Danjou is taken out of the crypt.

Sergent-chef LaBella described the symbolic meaning of the battle of Camerone:

the appeal of Camerone to a legionnaire is as natural as instinct. He reaches out to it in his own heart, because it is part of his own pain. It is the great reminder to the legionnaire that the sand is always blowing in his eyes, the battleground is always ill-chosen, the odds are too great, the cause insufficient to justify his death, and the tools at hand always the wrong ones. And, above all, nobody cares whether he wins or loses, lives or dies. Camerone gives the legionnaire strength to live with his despair. It reminds him that he cannot win, but it makes him feel that there is dignity in being a loser.

(in Bocca 1964: 6)

The inevitability of death is ever present in the life of the soldier, and particularly so for the legionnaire. American poet Alan Seegers, who met his end in 1916 fighting with the French Foreign Legion, wrote his poem *Rendezvous* which beautifully illustrates the mindset of these legionnaires:

But I've a rendezvous with Death
At midnight in some flaming town,
When Spring trips north again this year,
And I to my pledged word am true,
Shall not fail that rendezvous[3]

The legionnaire might not find meaning in fighting for the fatherland, but he does find meaning in a heroic death. This is born from a "supreme disdain for death. The composite legionnaire has all the sublime virtues brought out by war and displays virility and superiority" (Cooper 1969: 74) and this has been the strength of the Legion and the best guarantee of military performance.

Cut off from his past and isolated from reality, the legionnaire's life is worth very little to anyone, and least of all to France. This has inevitably affected morale, particularly during periods of peace and boredom when the Legion is given construction duties. Legionnaires are particularly sensitive to spouts of depression, which is commonly called *le cafard* after the cockroach that "metaphorically eats away at a man's brain, apparently devouring both mind and soul" (Gilbert 2010: 105). Legionnaire Frederic Martyn described the effects of *le cafard* as:

a form of mania supposed to be peculiar to the Legion – it is nothing more nor less, according to my idea of it, than a sort of hysteria set up by the action of a monotonous regime upon restless active natures in that climate. It is nature calling out insistently for change. It is but rarely that manifestations of the *cafard* end in tragedy; in ninety-nine cases out of a hundred they simply assume the character of an ordinary drunken quarrel or an extraordinary drunken spree.

(Martyn 1911: 223)

In other cases, however, depression and the oppressiveness of the Legion have led men to suicide, and once this avenue has been opened, it often spreads across the regiment. Patrick le Poer wrote in 1880 that when one man shoots himself an epidemic seems to set in; men hear every day in hut or tent or guard room the ill-omened report; soon they go about looking fearfully at one another, for no one knows but that he is looking into the eyes of a comrade who has made up his mind to die.

(Le Poer 2010: 90)

Suicide, like desertion, is harmful to a regiment. It reduces manpower, brings down morale and exposes a weakened army to the enemy. The legionnaire does not fear death; while he may run steadfastly to his death on the battlefield, he may also find very little to live for: fighting and dying, with no hope of recognition, for a cause that is not his own and for a country where he has no past and probably no future. This ambivalent relationship with death is detrimental to the image of the Legion. It undermines the reliability of the legionnaires whose unpredictable suicidal tendencies and perceived lack of patriotic allegiance can at any time jeopardise the mission.

Patriotism and foreignness

Legionnaires are often categorised as mercenaries: they are foreign, paid, albeit poorly, to go to war for a cause that is not their own, and are unruly and unreliable. The "outlaw image of the Legion, its racism, anti-Semitism and anti-intellectualism, its aggressive, hard-drinking and brother-crawling culture" (Porch 2010: 620) attracted a certain type of man, as did the active lifestyle and opportunity for battle. Despite their different nationalities, legionnaires often shared a common background and attitude that are equally prevalent in mercenary forces: they

> were denied any of the motivations commonly thought essential for modern fighting men – patriotism, a desire to defend family and homeland, the certainty that one's national cause is righteous and, lastly, the crutches of a language and national character, even of a shared sense of humor, so essential in carrying men over the rough spots.

(Ibid.: xx)

Consequently, the French government treated the Legion with suspicion, neglect and violence. Captain Morin de La Haye compared the legionnaires to a "human beast (who) shows his teeth and is brought to heel only through the use of fists" (Richard 1890).

Although the Legion is integrated into the French Army, the legionnaires have historically been profoundly anti-French and contemptuous of its armed forces. As the Legion is perceived to be for foreigners, legionnaires have frequently objected to the presence of French nationals in 'their' territory. The linguistic advantage of the French legionnaires has also contributed to this resentment. Anti-French hostility was voiced by American Jaime Salazar who complained that "the French wasted little time organising the inevitable *mafia francophone*, making the non-French further resent their exclusivity and arrogance. The Legion was supposed to be for foreigners, for Chrissake!" (Salazar 2005). Furthermore, as Germans have often been a majority within the Legion, this has encouraged Franco-German antagonism. Kevin Foster, a former Royal Navy seaman and legionnaire in the 1980s, stated that "the French thought themselves above everyone else. The common denominator in the Legion was that everyone hated the French" (Gilbert 2010: 69).

The image of the legionnaire was already viewed with suspicion because of its foreignness and infamous indiscipline. The high desertion rate of the Legion, however, exacerbated this impression because desertion is associated with mercenary units and a sign of indiscipline and lack of loyalty. Desertion is part of the tradition of being a legionnaire and the Legion still loses approximately 300 soldiers each year. Furthermore, legionnaires approached desertion "as a challenge, a gesture, a personal statement that was part of the process of being a legionnaire. The point was not that the desertion should succeed but that it should be dramatic" (Porch 2010: 622). This claim is corroborated by the surprising number of legionnaires who subsequently re-enlist in the Legion under another name. Nonetheless, desertion does not appear to have been detrimental to regimental performance: legionnaires tend to desert under the duress of boredom in stagnant postings rather than during mobile campaigns. Desertion, however, remains a sign of cowardice and undermines the perceived dependability of the Legion. Consequently, many French generals have argued that legionnaires could not be trusted in campaigns and might even prove dangerous to the regular French troops. The Legion was employed principally in the colonies for two reasons: first because French lives were judged to be too important to sacrifice to France's imperialist ambitions, and second, because the French government suspected the Legion's loyalties and therefore preferred to fight her national wars without foreigners. Whereas desertions are a common occasion in the Legion, rebellions against the fatherland remain few and far between.

The Legion's estrangement from the French Army is further amplified by its image as an elite force of mythical proportion. Its motto, *Legio Patria Nostra* (the Legion is our homeland), suggests that the legionnaires owe allegiance to the Legion, and not to the 'fatherland'. The regimental flags are decorated with the inscription *Honneur et Fidélité* (Honour and Loyalty) rather than *Honneur et Patrie* (Honour and Homeland), which adorns the flags of the regular army. This is a calculated and symbolic difference between the regular French soldiers and the legionnaires; the latter were, after all, mercenaries, and it would be pointless and insulting to ask them to fight for a 'fatherland' that was not their own. France recognised that the nation and patriotism were a priority in the legionnaire's paradigm and therefore made allowances for this army of foreigners: if the situation arose where a legionnaire came from a country with which France was at war, the soldier was asked whether he wanted to participate in the conflict. Legionnaire James Worden wrote in the 1950s that the Legion was introduced to recruits as *the only acceptable beneficiary* of a legionnaire's loyalty:

it is at this time that the almost fully trained recruit realises that, although on enlistment he swore an oath of allegiance to the French flag, that flag is wholly represented by the Legion, and only the Legion. He may be termed a soldier but he is not part of the French Army, and for him there will only exist the Legion and its officers. The French regular army is as remote as the man on the moon, and will not even exist in his mental make-up.

(Gilbert 2010: 29)

The Legion also accentuates its distinctness symbolically by its uniquely slow parade march and a uniform that does not sport the French flag.

On the other hand, the assumption that legionnaires are unpredictable and unreliable because of their foreignness is contradicted by the incredible courage that these men have repeatedly displayed. The Legion has been a useful tool in French foreign policy and has served its fatherland for the past 180 years. Loyalty and servitude to France are an intrinsic part of the training of the legionnaire. Article 1 of the *Code d'Honneur du Legionnaire*, which every recruit learns by heart in his own language as well as in French, defines the legionnaire as a volunteer who accepts the responsibility of serving France with honour and fidelity: "*Légionnaire, tu es un volontaire servant la France avec honneur et fidélité*" (Legionnaire, you are a volunteer, serving France with honour and fidelity). Furthermore, although the Legion is staffed with foreigners, it is very much a French institution, "tightly bound up with French prejudices and with French vanity" (Porch 2010: 632). France has adopted its legionnaires, awarding them – and their children – French nationality, conditionally and upon request, after three years of service. This concept of "*Français par le sang versé*" (French by the blood shed) was institutionalised in 1999. Despite ongoing prejudices against foreigners, mercenaries and the Legion, France's policy of naturalising its foreign combatants indicates that the legionnaires have earned the right to be French through their service to and sacrifice for the 'fatherland'.

Integration

Although the French Foreign Legion is a force made up of foreigners, and indeed mercenaries by most definitions, it is nonetheless an integral part of the French army. Full integration into the French armed forces required a record of military successes and evidence of reliability and state control that were progressively realised over the Legion's 180 years of service. The French government has always exerted control over the Legion by stipulating its recruitment policies, over-viewing training and setting the mandate for the legionnaires. At the same time, it is through its unusual autonomy and the nurturing of 'sacred rituals' that the Legion has developed such a high degree of unit loyalty among its men.

The Legion's officers are nearly all French officers who are serving in the *Armée de Terre* and have volunteered to spend some time in the Legion. Initially, this position was looked down upon and viewed as some sort of punishment. As the Legion earned itself a reputation as an elite and professional force, it began to attract the *crème* of the French officers. On the other hand, this brought about accusations that the Legion was draining the French Army of its best officers. Overall only 10 per cent of the Legion's officer corps has served as former legionnaires and *sous-officiers* (NCO). Officers are traditionally uninvolved in the day-to-day running of the regiments, which is the responsibility of the NCOs. Despite the presence of French officers, the Legion maintains its regimental autonomy and is allowed to administrate its disciplinary measures and training programmes at a local level.

Between its creation and its near-end in 1963, the Legion has hosted more than 600,000 legionnaires. In order to avoid the creation of a Praetorian Guard, however, any national component has to be kept at a maximum 25 per cent of the total. German speakers, including Swiss and Austrian nationals, have traditionally made up a comparative majority of legionnaires, but have never exceeded 35 per cent of the Legion's force. Although the Legion was closed off to Frenchmen until 2007, French nationals, under a declared identity, have also been an important pillar of the French Legion and today represent about 24 per cent of its soldiers. Camaraderie along national lines has been popular, encouraging cliques to form and parallel hierarchies to develop within a regiment, sometimes leading to disciplinary problems. Despite recruitment numbers decreasing recently in most Western armies, the Legion can afford to remain selective with an average of eight candidates for every opening compared to about three candidates per opening in the *Armée de Terre*.

Training

All recruits have to undergo a harsh training programme in the process of becoming a legionnaire. It is through training that the French government – or any government for that matter – forges the characters of the soldiers that the country needs. Basic training today takes place in Castelnaudary in the Languedoc-Roussillon, and is conducted by the 4ème Regiment Etranger. Over fifteen weeks, recruits are taught to adapt to the stringent requirements of the Legion whose objective is to create soldiers with a high level of stamina and the capacity to follow orders.

Unsurprisingly for a regiment of multinational soldiers, the first order of business is to make sure that all soldiers quickly reach a workable level of French. Each recruit is assigned a *binome*, a French-speaking partner to assist him during instruction. Languages other than French are prohibited, a policy that is implemented with extreme punitive measures. Each recruit is required to learn by heart the *Code d'Honneur* although he is also given a copy in his mother tongue to ensure that the meaning of the Code is well understood. The Legion's songs are also used for language instruction. Recruits are exposed to the French language through the songs that are a cornerstone of the Legion. Learning French not only serves to communicate; it also creates *esprit de corps* and contributes to forming an emotional and cultural attachment between the legionnaire and the fatherland:

the Legion authorities regard the melodious, rhythmic, vigorous, and unanimous execution of these songs by detachments on the march as of the first importance. It is considered that singing in chorus does a great deal towards the inculcation of an esprit de corps in men who have no other universal way of communicating together.

(Gilbert 2010: 43)

Central to the training process is the destruction of the individual and his rebirth as a legionnaire. Basic training was therefore more a psychological process than a physical exercise. The focus on cleanliness and discipline rather than small-arms training and technical instruction was bewildering and strenuous for the recruits itching for action. Recruits had to first learn the Legion way of doing things:

it may not be the most logical way or the simplest way, it may seem like the most stupid, ridiculous method in the world – but it is done that way and you are going to do it that way – even if it takes all night and all the next day.

(Jameson 2012)

Pádraig O'Keeffe, describing his first few weeks in the Legion, explained that

the Farm was all about your mentality. It wasn't about whether you could make it physically as a soldier. It was about discovering weaknesses and identifying those recruits who would invariably seek the easy way out. The Farm instilled a kind of attitude, a Legion outlook on how things should be done and why you never, ever quit (…). It was a combination of physical exhaustion, mental fatigue, self-doubt and, for some, probably even a bit of fear (…). There is only one word that can summarise the experience – hell. The idea of the Farm, as one instructor put it, was to "take you down to the basics and then build you up". Nothing was written down, but at the Farm you'd get everyone at the same level, starting with lessons in how to wash, and then work upwards.

(O'Keeffe with Riegel 2007: 51)

Captain Morin de la Haye explained already in 1886 that it was necessary for the training system to be

pushed to its maximum intensity. In the beginning this system produces vague impulses of revolt. Refusal to obey, desertion, cases for court martial occur daily; men shot at their officers, others look to unhorse them; in the Legion, the officer must expect everything.

(cited in Richard 1890: 132)

The rigorous and at times extreme training programme was also a method of filtering the recruits:

at times we carried out forced marches. We leave for the south. Everyone must keep up. The man who remains in the rear risks death by starvation or being taken by Arab dissidents. After these marches, the number of stragglers is considerable: one must be strong to endure. This is the Darwinist survival of the fittest applied to the troops.

(Ibid.)

Sleep deprivation, food deprivation and physical torture are used to break down the recruit. The methods produce formidable results, however: de la Haye concluded that "in several months, one has a solid troop, manoeuvrable, admirably practiced in shooting and marching, and entirely in the hands of its leaders" (ibid.).

The rigours of basic training, the stringent disciplinary measures, along with the traditions of the Legion that are imbued on the recruit from day one, contribute to forging a common experience of what it means to be a legionnaire. Extreme hardship and shared lifestyles created an *esprit de corps* that became part of the identity of the legionnaire. Those who successfully complete the 120 km *Marche Képi Blanc* in week five and survived the duress and harsh discipline of the next 10 weeks were then assigned to one of the Legion's eleven regiments where they remained for the duration of their contract. Professionalisation and specialisation mostly took place within these regiments.

Discipline and control

Given the heterogeneity of the legionnaires who come from all sorts of backgrounds – from criminals and deserters to intellectuals and princes — the Legion has found it necessary to use draconian measures in order to instil discipline and impose the authority of the NCOs in charge. In theory, the *règlement de discipline* is the same in the Legion as in the rest of the *Armée de Terre* with the difference that legionnaires need to seek permission before owning a vehicle or getting married. In practice, however, the remoteness of the garrisons and the legionnaires' high rate of desertion have created both an opportunity and a necessity for NCOs to exercise unusual autonomy in ways of disciplining their troops.

Where training fails to bring the legionnaire up to scratch, disciplinary measures have been used both as punishment and as a means to ensure *esprit de corps*: "unit cohesion was guaranteed by fear of punishments long outlawed in Europe, if indeed they had ever existed there" (Porch 2010: xx). Traditionally, the Legion favoured "short, sharp punishments (that) were in effect more humane and more efficient than a prolonged jail sentence, which deprived a unit of a man's services and stained his service record" (ibid.: 625). Some of the more brutal punishments included *la crapaudine*, *la pelote*, and *le silo*. *La crapaudine* was described by Frederic Martyn in 1892: after an Italian legionnaire struck his sergeant, he was

stripped naked, his hands were pinioned behind his back, and his ankles tied together. Then his ankles were lashed to his wrists, and he was thrown on the ground looking very much like a trussed fowl. The agony incidental to this constrained position must have been almost beyond human endurance after a time; but in this poor man's case the punishment was intensified by the fact that in no long time after he was tied up his body was literally covered with a swarm of black ants.

(Martyn 1911: 186)

The *silo* was used in the nineteenth century to punish serious crimes. It

consisted of a funnel-shaped hole in the ground, broad at the top and pointed towards the bottom. A regular funnel. Into this hole, used as a cell for solitary confinement, the misdoer would be thrown, clad only in a thin suit of fatigue clothes, without a blanket or any protection at all against the rain or against sun, at the mercy of the heat by day and the cold by night. (...) They very soon became ill from the foul vapours. When at length they were taken out of the silo, they could neither walk nor stand and had to be carried into hospital. Now and then a silo prisoner died in the hole.

(Rosen 2010: 228)

The *pelote* probably remains the most popular form of punishment: perpetrators have their packs filled with stones or sand and are forced to march in circles and do press-ups for hours. Legionnaire Jacques Weygand described the back-breaking effects of the *pelote*:

the braces cut into the men's shoulders; elbows and knees are bruised or bleeding from repeated contact with the stony ground. No matter; chin-strap braced, watch in hand, implacable, the sergeant threshes his human grain. Sweat runs in dark runnels down faces that are clotted with dust; and a lesson never to be forgotten is stamped into the most rebellious spirits.

(Gilbert 2010: 62)

Generally however, punishment and corrections take the form of a swift beating attributed by the sergeant or corporal. NCOs enjoy an elevated status in the Legion and are given more responsibility than in most other armies. They are in effect the backbone of the Legion. On the other hand, the unique hierarchical system of the Legion has made its NCOs "virtually unaccountable for their actions, least of all to French officers, who traditionally held aloof from the day-to-day running of their units" (Porch 2010: 625). NCOs therefore have often earned themselves a reputation as brutes, or even sadists. Attempts to appeal against unfair or unreasonable treatments were usually discouraged through violent means. Excessive brutality, however, can be dangerous, especially when dealing with legionnaires, as unpopular NCOs could quickly find themselves with a bullet in the back of their head. Authority and respect were key, and overall the Legion has arguably maintained a reputation for being hard but just. Cruelty has been justified as necessary to harden the recruits.

Finally, despite the high level of autonomy that is awarded to the NCOs, the French government maintains absolute power over the Foreign Legion:

> the Legion never fully trusted its soldiers, an extension of French suspicion of foreigners that caused the country to place them in a separate corps in the first place, and took elaborate steps from the 1920s to create a secret service whose primary mission was to oversee the loyalties of its own troops.
>
> (Ibid.: 623)

Today, every new recruit is screened to ensure that he has no record of having committed a serious crime. Furthermore, at least 90 per cent of the officer corps is made up of Frenchmen who exert direct control over the units, even though they often delegate minor tasks and training to their under-officers. The Legion is entirely reliant on the French government for logistics, including housing, equipment, ammunition, food, transportation and backup support. France can also order the arrest and imprisonment of legionnaires who transgress or attempt to desert the army. The fatherland maintains the right and the ability to punish its soldiers, and may even order the dissolution of a regiment. Attempts to challenge the French government have shown the latter's ultimate control over the Foreign Legion: President Charles de Gaulle disbanded the 1ᵉʳ REP in 1962, cut the Legion's numbers in half, and was only dissuaded from dismissing the Legion entirely by the personal appeals of Pierre Messmer, the French Minister of War and a former legionnaire.

Loyalty

Although the French may view the Legion with suspicion, the legionnaires have demonstrated an extraordinary level of unit cohesion. Loyalty to the Legion is developed through the common experience of rigorous training, but also through the sacred rituals and traditions that lend this unit a unique identity of which the legionnaires can be proud. 'Cultural' indoctrination is an essential part of building the legionnaire. From the moment they arrive in Castelnaudary, recruits are exposed to the 'unique' attributes that separate the Legion from the rest of the French Army.

The Legion aspires to be more than just a military force: its soldiers do not fight for France; they fight for glory in its purest tradition. The experience of becoming a legionnaire and the life that he leads while under contract "will not appeal to anyone who does not love the soldiering trade for its own sake" (Martyn 1911: 86). The Legion presents itself as an elite, professional force, and the legionnaires must demonstrate a true military vocation in order to survive the five plus year of their enlistment. Legionnaire Christian Jennings outlined the particular military attributes of the Legion:

> over the decades the Legion had developed its own kind of soldiering. It was not soldiering where all hung on victory or defeat, as in other armies,

but rather a stylish profession of arms, aimed at bringing greater glory to France and to the Legion.

(Jennings 1991: 27)

The mythical image of the Legion has also been improved by cinema and literature that have portrayed legionnaires as legendary combatants and glorified the Legion as the ultimate warrior creed.

The men who join the Legion as volunteers are entering a community with a long history and a strong sense of tradition. The Legion offers a second chance to its men, and defends their anonymity with zeal: "in joining the Legion, men do not join only an organised, disciplined institution; they enter a situation which is more or less recognised by the whole world and sanctified by its sacrifice. They belong – many for the first time" (Cooper 1969: 74). Loyalty in the Legion is both top-down and horizontal:

> from the highest to the lowest, there is a strong bond among legionnaires which is stronger than any solidarity of rank. A private in the Legion knows that his colonel prefers him to any officer who is not in the Legion. The Legion is like a large family that needs only itself to exist.
>
> (Bruckberger 1952: 30)

As part of the *Code d'Honneur*, the legionnaires vow to fight for their brothers in arms, regardless of their background, and to demonstrate the absolute solidarity that one should feel for a member of their family: Article 2 — "*Chaque légionnaire est ton frère d'arme quelle que soit sa nationalité, sa race, sa religion. Tu lui manifestes toujours la solidarité étroite qui doit unir les membres d'une même famille*" (Each legionnaire is your brother in arms whatever his nationality, his race or his religion might be. You show to him the same close solidarity that links the members of the same family). This strong sense of loyalty permeates the harshness of military life and welds the legionnaires to one another and to their NCO.

The Legion maintains its difference from the rest of the French Army by nurturing its own cultural legacy: regimental mottoes focus on honour and fidelity towards the Legion, not towards the French Army. Legionnaires are visually distinguished from their French comrades by their uniforms and the *Képis Blancs* recruits are expected to learn 52 traditional songs of the Legion, many of which narrate the glorious history and harsh challenges that their predecessors encountered. The annual celebration of the French Revolution on 14 July further demonstrates the differences between the Legion and the French Army: the legionnaire march at the slow pace of 88 steps per minute, compared to the 120 steps that are the requirement for all other units; because the Legion must never be separated, the legionnaires do not break ranks when they reach the French President's chair, but turn in one movement as the head of the army stands to salute them. The mystique of the Legion is recognised by the French public who enthusiastically applauds their legionnaires as they march down the Champs

Elysées. This contributes to instilling a sense of pride and belonging in these stateless combatants.

The *esprit de corps* that is developed through training and traditions is cemented by the camaraderie that inevitably grows from shared experiences. The third article of the *Code d'Honneur* stresses the importance of traditions and camaraderie: "*Respectueux des traditions, attaché à tes chefs, la discipline et la camaraderie sont ta force, le courage et la loyauté tes vertus*" (You respect your traditions and your superiors. Discipline and friendship are your strengths. Courage and honesty are your virtues). This value is extended to the battlefield, where each legionnaire is bound by his oath to never abandon his dead or wounded comrades: Article 7 – "*Au combat, tu agis sans passion et sans haine, tu respectes les ennemis vaincus, tu n'abandonnes jamais ni tes morts, ni tes blessés, ni tes armes*" (In combat you act without passion or hatred. You respect vanquished enemies. You never surrender your dead, your wounded, or your weapons). Camaraderie, loyalty and the refusal to surrender have at times led the legionnaires straight to their deaths, as evidenced in the battles of Camerone and Dien Bien Phu. Nonetheless, this extreme devotion and blind obedience to the Legion have resulted in an impressive track record of combat performance.

Conclusion: lessons learnt

In 180 years of history, the French Foreign Legion has overcome social norms and prejudices towards mercenaries, and successfully demonstrated to the world that a foreign and heterogeneous regiment can create excellent and reliable soldiers. This is a direct contradiction to the convention that conscript soldiers make the best combatants because they have the right motivations and are easily controlled by the state. The successful integration of the Foreign Legion into the *Armée de Terre* offers an opportunity to explore shifting norms and adaptation techniques that can inform the debate on the cohabitation of foreign and private combatants and conventional armed forces.

The men who enlist in the Foreign Legion are without exception volunteers. They join the Legion out of their own free will and, unlike the stereotypical mercenary, they do not do so for money – there is none to be made in the Legion. Their motivation varies from vocational aspirations to existential pursuits and while legionnaires are not known for their patriotism, their loyalty to the Legion and among themselves is legendary and has effectively spirited their performance on the battlefield. This loyalty has been deliberately instilled through a programme of ideological and cultural indoctrination and by exposing the legionnaires to extreme situations where they have had to create fraternal relationships in order to survive emotionally and physically.

The physical training, with its focus on stamina and obedience, along with the strong ties that unite the legionnaires, have effectively produced a force that is both effective and reliable. Despite an uncertain start, the Foreign Legion has been a pillar of French defence and an asset to French foreign policy, enabling her imperial ambitions and upholding the reputation of a French Army whose record in the last two centuries has been uneven. It has also been a deliberate domestic policy to deal with the problems of refugees and immigrants: "together with ridding France of some of her foreigners, the Legion has policed her colonies and fought her wars, while at the same time earning fame as a crack unit with an international reputation for military efficiency" (Porch 2010: 630).

On the other hand, the Legion has also experienced many non-military obstacles that have marred its image as an elite force. Despite evidence that foreigners are not necessarily less reliable than conscript soldiers, legionnaires have suffered from prejudices and accusations of mercenarism. This perception has been exacerbated by a high rate of desertion, and by allegations of racism, anti-Semitism and brutality that have been associated with the Legion since its participation in the *mission civilisatrice*, particularly in Algeria. The questionable backgrounds of many legionnaires and the hard-drinking, violent culture of the regiments have also contributed to the controversy surrounding this unusual military unit. "The cohesiveness of the Legion, its racism, its elite combat image, the preference of its commanders for strong-arm solutions to problems of colonial agitation and its recruitment made it especially useful to support a shoot-first policy" leading to accusations of over-zealousness, particularly in the Kolwezi operation where legionnaire "were trigger-happy and killed far more people than necessary" (ibid.: 623).

Furthermore, the polyglot nature of the Legion made it at times difficult to manoeuvre, undermining its efficiency in battles. This was exacerbated by France's neglect, evidenced in the parsimonious approach to equipping its foreign troops: "the real injustice of Legion service was that good troops too often failed to realise their true potential because they were inadequately trained, armed, led and supplied" (ibid.: 624). The difficulties that the Legion has faced is testimony to its isolation from the rest of the French Army where integration has yet to be accepted fully – although the situation appears to have improved since the reform of the Legion in 1962.

Overall, the French Foreign Legion has survived two centuries of military history, and it does not appear that it will be dwindling in the near future. Its military successes, long-standing loyalty and position in the eyes of French society have undeniably proven that nationality and patriotism are not correlated with military performance. Indeed, it appears that the most important concern in society relates to the state's control over the armed forces, not to the origin of the combatants. Despite its relative autonomy, the Legion is led by French officers and remains entirely dependent on the French government, which can at any time legally and effectively hold the legionnaires accountable for their actions.

The Legion has successful adapted into its new role in the twenty-first century. Current geopolitical trends indicate that, in the near future, Western armies will be deployed in peacekeeping and nation-building operations rather than in straightforward combat roles. As a participant in multinational operations, the Foreign Legion has represented the fatherland and successfully cohabitated with other French troops, even adapting its traditional uniforms to fit in with the UN's requirements. The Legion has been fully integrated into the

French Army since its first mandate in Algeria, and has supported the French in her wars of national defence and colonial expansion. Notwithstanding accusations of mercenarism, the Legion remains under the authority of the French government, and is a recognised and celebrated component of the French *Armée de Terre*. The French Foreign Legion is a success story that shows that mercenaries can successfully be integrated into a nation's foreign policy and can cohabite harmoniously with the national armed forces.

Notes

1 As exemplified by the Manifeste des 121 of September 1960 signed by intellectuals including Jean-Paul Sartre, Simone de Beauvoir and Andre Breton, among others.
2 Interview with a former legionnaire, Paris, June 2012.
3 Alan Seeger and William Archer, *Poems* (New York,: C. Scribner's Sons, 1917).

5   Executive Outcomes

Nowhere have mercenaries and private military and security companies been more active – or more controversial – than in Africa. Ever since the 1960s, select African countries have been plagued by mercenaries trying to turn a profit from the post-colonial conflicts raging across the continent. The attempted coups against the governments of the Comoros Islands, the Seychelles and Equatorial Guinea that were orchestrated by the mercenaries Bob Denard, Mike Hoare and Simon Mann, respectively, further exacerbated the tensions between non-state combatants and political leaders. On the other hand, the fall of the Soviet Union brought an end to Western and Soviet subsidies and created a new market for foreign military support, particularly in weak and war-torn African states. The 1990s therefore saw the beginnings of a new military model in Africa: the privatisation and commercialisation of mercenary troops by Western and South African entrepreneurs.

Private military companies (PMCs) have become a de facto military solution to the national and personal insecurities of many African leaders whose states suffer from weak institutions and inadequate military capabilities. These states are particularly vulnerable to the institutional disruptions and civil–military rivalries that result from outsourcing military services to foreign private military companies. This chapter starts by outlining Africa's contentious military history and its experience with mercenaries and private military companies. By using Angola as a case study, it subsequently describes the particular attributes and traditions of the national armed forces in a weak and conflicted state and highlights the impact that hiring PMCs has on the identity and military effectiveness of the national army. Despite strict anti-mercenary norms, the African model of privatisation appears to be unique in its political and legal freedom to provide select military services. The successes and failures of this security model are explored in a wider context of military traditions and show that while privatisation might appear as a short-term solution to many problems plaguing African leaders, it is fraught with difficulties and can ultimately be damaging to the identity, loyalty and hierarchical structure of the national armed forces.

Africa's armies

Africa in the second half of the twentieth century was the playing field for mercenaries searching for adventure and easy money. Following the process of decolonisation, many states in Africa were left with weak political and military institutions. A divided continent saw no less than 85 successful military coups and 19 presidential assassinations since the early 1950s (Kieh and Agbese 2004: 45). Out of 54 independent African states, at least 35 countries have at one time or another suffered from civil war or inter-state war. The conflict-ridden continent therefore provided many opportunities for soldiers of fortune to tempt their luck and find employment. The flow of foreign mercenaries into Africa created many problems, particularly for the national armed forces.

The pre-colonial military traditions and the legacy of colonialism significantly affected the military identity of Africa's soldiers. Critics argue that African armies "are unprofessional, lacking both technical expertise for combat and political responsibility to the state" (Howe 2001: 9). Furthermore, African leaders have shown a tendency to favour "military loyalty at the expense of capability" and deliberately "debilitated their own security forces in an effort to preserve political power" (ibid.: 2). Consequently, soldiers in many African states have a very different socio-political and military environment in which to manoeuvre compared to their European and American counterparts. This has inevitably affected their relations with the foreign mercenaries who are hired, either by the state or third parties, to wage war on their territory.

Pre-colonial armies

Prior to the eighteenth century, there was no history of a structured standing army in Africa. Apart from Northern Africa and arguably West Africa, the continent's vast territorial expanse was not conducive to competition between the different tribes as the availability of space enabled them to move away when faced with an armed threat. There was therefore no urgency to develop an organised system of defence. This contrasts with the European continent where the rapid development of military organisation and technology was the result of a constant struggle for food and territory that consumed the different tribes living within the very restrained geographic space. Herbert Howe argues that the "fear of invasion has traditionally encouraged states to develop professional militaries: the Napoleonic Wars accelerated Europe's military capabilities and, specifically, the creation of a professional officer corps" (Howe 2001: 32). Although inter-tribal violence was present in pre-colonial Africa, these wars were neither as frequent nor as deadly as their equivalent in Europe.

Most pre-colonial 'states' were conquest states with a fused culture of military and political institutions. African tribes generally adopted a warrior tradition described by Ali Al'Amin Mazrui as a "sub-system of values and institutionalised expectations which define the military role of the individual in the defence of his society, the martial criteria of adulthood, and the symbolic obligations of manhood in time of political and military stress" (Mazrui 1977: 2). Military service was considered to be a transformative experience that distinguished the men from the boys and marked their active participation in the tribe's political organisation, with the objective of ensuring its survival. Political leadership was consequently derived from military prowess: "the king and commander-in-chief was the strongest and bravest warrior of them all" (Uzoigwe in ibid.: 31).

The shift from the 'warrior-politician' to the 'professional warrior' occurred in the eighteenth century with the development of military professionalism in the Zulu Kingdom: the Zulu leader, Shaka, introduced drilling, marches and logistical organisation to his army. Shaka's military reforms were driven by European encroachment into African territory. This created

> the need (for Shaka and others) to raise an army of a size beyond the capacity of a single man to control and led to the creation of war chieftaincies; the improvement of weapons and the introduction of new ones required skilled and experienced men to handle them, and thus a class of professional, or full-time, and semi-professional warriors rose.
>
> (Smith 1981)

This trend spread throughout central and east Africa as the warrior institutions and values dictated by Shaka were adopted by neighbouring kingdoms. The change was drastic: the warrior was no longer an integrated political citizen but segregated from the state and organised into a powerful force capable of challenging his king. The Zulu Wars in 1879, a good 50 years after the death of Shaka, demonstrated the effectiveness of the military reforms spearheaded by the former king and carried on by his nephew, Cetshwayo: the Zulus defeated the British at the Battle of Isandlwana on 22 January. The victory was short-lived, however, as the British, with their superior weaponry, crushed the Zulus at the Battle of Ulundi just six months later.

The colonial military legacy

The colonial occupation of Africa further changed the dynamics of the armed forces. The mandate of the colonial armies was to protect the national boundaries and suppress domestic unrest, and they were used by the colonial powers as an instrument of control and coercion: "during the colonial era, the mission of the military was clearly specified: it was to facilitate the exploitation of African resources for the benefit of Europeans. It was used principally as an instrument of domestic repression by colonial officials" (Kieh and Agbese 2004: 190).

Colonial armies were entirely staffed by European officers who "rarely, if ever, crossed the civil–military dividing line by challenging civilian rule. Officers, especially from Britain and France, were steeped in the tradition of military acceptance of civilian control" (Howe 2001: 28). These armies were used by the major colonial powers to impose and ultimately to maintain their rule on the territory, turning the "colonial military (into) merely the armed wing of the colonial state" (Kieh and Agbese 2004: 190). Instead of developing a local police force, the military was repeatedly used for policing missions, setting a precedence for future armies to interfere in civilian disputes. Although colonial military forces "displayed some professional traits of political responsibility and military capability", they failed to "develop an indigenous and professional officer corps; the forces laid the groundwork for future unprofessional militaries" (Howe 2001: 28). The colonial powers felt threatened by the possibility of military retaliation, and consequently declined to develop a competent indigenous officer corps or even a professional soldier: "the ideal colonial soldier was supposed to be illiterate, uncontaminated by mission education, from a remote area, physically tough, and politically unsophisticated" (ibid.: 33). The absence of external threats further discouraged any urgency to train and equip the army, leading to an inexperienced, untrained, under-equipped army, with no substantial leadership at the end of the colonial period.

The transition from colonialism to independence was largely peaceful, leaving "the armed forces on the political sidelines, neither needed by the metropolitan power to suppress nationalist movements nor countered by guerrilla or liberation armies sponsored by indigenous politicians" (Houngnikpo 2010: 57). The national army in post-colonial states was perceived as a colonial and collaborationist institution, and regarded with fear and suspicion by the newly independent governments:

> nationalist leaders saw them as remnants of imperial rule. Though they had won glory by serving overseas in the two World Wars, their imperial 'credentials' caused them to be regarded in some quarters as armies of occupation or at best as mercenaries in the service of a foreign power.
>
> (Ibid.: 56)

The national armed force, therefore, have had to adapt to the requirements of the new African leadership in order to maintain a (reasonably) healthy budget and a position of relative authority and influence. The civilian government of many post-colonial states believe that "the function of the military was to defend authority, and not society; it almost always came to the support of the state in suppressing political dissidents, democrats, socialists and others" (Perlmutter 1981). Consequently, the armed forces have repeatedly been used for domestic deployment to suppress civilian uprisings and act as a police force. The independent African state is frequently accused of being the reincarnation of the colonial system, catering its policies towards personal enrichment and ignoring the needs of its citizens.

Military coups and intervention

Military intervention in African politics has been prevalent since the end of the colonial period. As mentioned above, in the last 40 years, the continent has experienced at least 85 military coups, of which five have resulted in the death of the incumbent president. Interference of the armed forces in the political affairs of the countries is known as 'praetorianism', which "characterises a situation where the military of a given society exercises independent political power within it by virtue of an actual or threatened use of force" (Perlmutter 1977: 89). Praetorianism dates back to the Roman Empire, when an elite unit of professional soldiers was recruited to defend the legitimacy of the empire against rebellious military garrisons. Known as the Praetorian Guard, this military force was the only armed presence allowed into the city of Rome, and as such exerted huge influence on the candidature of the Senate: "thus it was able to manipulate a widely subscribed concept of legitimacy and to attain a degree of influence disproportionate to its size and military resource" (Perlmutter 1977: 9). In the modern era, intervention in the political arena is to be expected if the state lacks legitimacy in the eyes of the armed forces: indeed, "when civilian politicians rigged an election and defy civil society, it is difficult to demand military subservience to an illegitimate regime" (Houngnikpo 2010: 59). Unfortunately, this appeared to be the case in most post-colonial states, leaving many African politicians fearful for their jobs.

An alleged lack of political legitimacy has in fact been the main trigger for the military's interference in the affairs of the state: "military intervention in African politics was rationalised by the soldiers as a patriotic and selfless exercise to rid the continent of corrupt, inefficient, incompetent and decadent politicians" (Kieh 1992: 5). Armies displayed a 'messianic complex' and viewed themselves

> as the saviour of their countries. The members of the armed forces would routinely argue that besides the military, no other section of the society is capable of liberating their countries from maladministration and chaos, which are usually the handiwork of civilian institutions.
>
> (Kieh and Agbese 2004: 41)

This is particularly the case in weak states where a lack of rule of law and institutional capability of the government provide the military with an opportunity to grab power: according to Mathurin Houngnikpo,

> the military's subordination as hypothesised by Samuel Huntington has not worked very well on the continent, because most African states tend not to operate within an established framework of viable and widely based institutions, even when they have been legitimised.

Samuel Finer explains in *The Man on Horseback* that in many developing countries the organisational capability of the army far surpasses civilian institutions: "even the most poorly organised or maintained of such armies is far more highly and tightly structured than any civilian group" (Finer 2002: 6).

The reasons why the armed forces do not always interfere in politics, however, is that they are technically unable to "administer any but the most primitive community" and "lack legitimacy: that is to say, [they have a] lack of a moral title to rule" (ibid.: 14). The military require moral authority to justify their use of violence, and often prop up a puppet government to conceal their activities and receive a stamp of legitimacy after a military coup – this was the case in Nigeria in 1979 and 1993 when the Nigerian Army briefly transferred power to a civilian government over which it maintained power. States that have a developed political culture are less vulnerable to military intervention because of the normative expectation that the governing institutions must be civilian. A history of military interference in domestic politics has left its imprint on the self-perception and identity of the armed forces. Politicians lacking confidence in the legitimacy and longevity of their rule have attempted to reign in their unruly and unreliable armies.

Weak military institutions

Threatened by the prospect of a coup d'état, African leaders have repeatedly sought to weaken the military branch of the state. Rulers who fear the power of the armed forces intentionally enact policies to curtail the operational capability of their armies. President Mobutu in Zaire, for example, deliberately weakened his army by encouraging jealousy and suspicion between the different units and the officers. By playing them off against one another, the ruler could avoid an officer's coup and used informants to anticipate any plots: "they removed officers who appeared untrustworthy (or too competent), played the regular militaries off against other armed groups, such as guards or mercenaries, and placed specific constraints on operational capabilities" (Howe 2001: 50). Although the position of the president was consequently safeguarded against a military coup, by deliberately encouraging dissent within the army, the ruler incapacitated decision making and the implementation of progressive policies. In extreme cases, the competition between officers also led to a civil war, as was the case in Liberia in the 1980s. In some countries, the weakness of the armed forces facilitated the spread of rebel groups who opposed the government and accused the military of corruption and incompetence: "in Sierra Leone's case, a weak RSLMF would later facilitate the growth of the Revolutionary United Front (RUF) through the RSLMF's technical incompetence and its absence of strong loyalty to the state" (Howe 2001: 37).

Furthermore, highly corrupt states often failed to pay their soldiers, whose salaries were progressively siphoned off by dishonest superiors and other intermediaries. The failure to pay soldiers forces the latter to bully the civilian population to gain access to their basic resources such as food, clothing and ammunition. It also pushes the military into pursuing commercial activities including smuggling and trafficking, which destabilises the local economy and can lead the country into a state of prolonged conflict, as was the case in Angola and the D. R. Congo in the 1990s. In addition, certain governments even encouraged the lower ranks in the army to extort money from its citizens as a substitute form of payment. Nowhere was this more prevalent than in Mobutu's Zaire, where his six private armies were given free range to loot the country. This policy alienates the civilian population, who lose trust both in the armed forces and in the civilian government. It is natural, therefore, that the second most cited justification for attempted coups is to preserve the interest of the military institution against the predatory and dangerous policies of the state. Armies rely on the government for their budget allocation, and any attack on their status, norms, identity and financial or physical well-being can trigger a military response.

The inability to count on the loyalty of their army, however, has compelled rulers to create parallel security forces and combat militias, often representing their own ethnic group. In multi-ethnic countries, the systematic recruitment by ethnicity of military personnel has led to allegations of favouritism and discrimination. "In some African countries, the armed forces are not national institutions (but) are dominated by people from particular regions or specific ethnic groups" (Kieh and Agbese 2004: 191), and consequently they do not always feel represented by the government in power. This has led to ethnic factionalism between the state and its military arm. In extreme cases, the armed forces deliberately target and repress select ethnic groups within the country, inducing fear among the population who retaliate against other civilians of the same ethnic group as the perpetrators. Howe argues that the targeting of civilians by the military "lowers political acceptance of the state by groups excluded from the military, (raising fears) of state-sponsored repression" (Howe 2001: 29).

The use of parallel forces to compensate for an unprofessional army – often a direct cause of government policy – is also a threat to the identity, professionalism and existence of the national army: "parallel forces angered existing national forces because of the presidents' implicit vote of no confidence, the groups' freedom from the military's normal chain of command, and their first choice in equipment" (ibid.: 45). This further affected civil–military relations by complicating the channels of communication between the military and the government. The lack of mutual trust and support between these actors challenges the authority, credibility and legitimacy of civilian rule over the military. The armed forces effectively loses respect for and belief in the government of their country, and reacts by taking decisions into their own hands and ignoring the orders of the civilian government. It also frequently leads to internal fighting between the military and the parallel forces. Nonetheless, it should be noted that parallel forces have repeatedly been perceived by the international community as being more professional than the regular army, and they provide the ruler with a respectable amount of personal security and stability by anticipating and preventing military coups against the regime.

Breakdown of military identity

The constant intrusions of the government and the loss of faith and support from the population can lead to a real crisis of identity for the armed forces. The army is a symbol of sovereignty, and "a purposive instrument (...); it comes into being by fiat (...) and is rationally conceived to fulfil certain objects" (Finer 2002: 7).

As an institution, the military's role is to protect society from external threats, and ensure internal stability. One of the first acts of a new state or government is usually to ensure that it has a functioning or semi-functioning army. When this army is unable to carry out its mandate to protect the population and the nation against domestic threats, either because it is being manipulated by the government or because it is too weak to defend the country against rebels, it runs the risk of losing its *raison d'être*.

As the budget of the national armed forces depends on the system of taxation in which the population participates, good civil–military relations are essential for the continuity and operationability of the armed forces. If it loses popular support, the army becomes isolated within the state. Its identity shifts from that of protector of the people to being an antagonist of the state. The nickname 'sobels' (soldier/rebel) was given to Sierra Leone's armed forces between 1991 and 2002 and perfectly illustrates the breakdown of the military's identity. When the army loses its status in society, it ceases to work as a national institution, thus forfeiting its role and its funding. The loss of identity and purpose leads to the demise of the military.

Finally, where civilian institutions are too weak to monitor their armed forces, a vacuum of authority arises. With no punitive measures or legal infrastructure, the state forfeits its authority over an army that it is unable to control. Jakkie Cilliers argues that "civilian control over the armed forces is the end result of a complex interaction among various factors, including formal legal controls, a strong civil society, and the nature of the armed forces" (Cilliers and Mills 1995: 8). As Africa lacks the "professional culture that acknowledges the supremacy of civilian and parliamentary authority over the military, the prospects for civilian control over the armed forces are not bright" (ibid.). Rampant poverty and a tradition of paternalism and patronage have also encouraged officers to pursue a military career as a means to political and or economic power rather than for patriotic or vocational reasons. Africa's armies have consequently suffered from a dearth of professionalism, compounded by the state's lack of credibility. This has led to a vicious circle of suspicions and sabotage in an environment already prone to conflict and insecurity. Political leaders have responded to this crisis by calling forth military support from their former colonial powers or from private security providers.

Africa's mercenaries

If the state's relationship with professional armies has been bad, Africa's long and dark history with mercenaries is even worse. The weak political and military institutions of many African states have attracted soldiers of fortune and encouraged domestic politicians, private individuals and foreign governments to interfere illicitly in the country's politics. Since the 1960s, Africa has experienced three models of mercenary intervention: (1) *ad hoc* regiments of independent mercenaries recruited for a particular strategic purpose and operating under the command of a mercenary leader. This type of intervention is exemplified in

Mike Hoare's missions on behalf of Katanga, a breakaway province of Congo, and later in his military operations against the Simba rebels in Congo. (2) Mercenary coups, organised and financed by a (usually foreign) benefactor and assembling a group of highly trained mercenaries to topple a regime and spearhead an uprising against the incumbent ruler. This strategy was made popular by Frederick Forsyth's 1974 novel *The Dogs of War* which depicts a company of European mercenaries hired to depose the president of a corrupt African country. This exact scenario was re-enacted in Equatorial Guinea in 2004, where the operation turned out to be a complete failure. (3) Trained military professionals employed for a short term by a foreign company and outsourced to a country to provide a temporary service of military consulting and potentially organising and leading combat operations against the domestic enemies of the contracting state. Private military companies, such as Executive Outcomes (EO) and Sandline International, were used respectively by Angola and Sierra Leone in the 1990s to counter the advance of armed groups of rebels. Mercenaries have come to Africa to topple regimes, replace soldiers, advise politicians and essentially fight in other people's wars. Africa's controversial experiences with mercenaries have informed the continent's strict anti-mercenary norms, although these have been practised inconsistently: despite legislation condemning the use of mercenaries, African leaders have increasingly turned towards private military and security companies to compensate for their own security failures.

Secessionist states

The end of the colonial system left behind an amalgamation of ethnically and religiously heterogeneous states. Sovereignty was conveyed *de jure* to the foreign colonies by an international community of Western states. Post-colonial states were thus empowered with new political and economic responsibilities "irrespective of their effective capacity to control their populations and territory and to fend off challenges from other states" (Englebert 2009: 5). This privilege was engraved in the United Nations General Assembly Resolution 1514 (XV) of 1960 which unilaterally and unconditionally conveys sovereign recognition to "all colonial countries and peoples". A legal "right to complete independence and the integrity of their national territory" set a precedent that condemned secessionist attempts from these newly formed states.

Nonetheless, after the departure of the foreign powers, secessionist movements began to claim ethno-linguistic, cultural and religious differences as grounds to separate from the rest of the country. Since independence, at least 10 of sub-Saharan Africa's 48 states have experienced a secessionist attempt, although nearly all of these have failed. Secessionist attempts took place in Katanga and South Kasai in D. R. Congo between 1960 and 1963, in Nigeria's province of Biafra between 1967 and 1970, in Senegal's province of Casamance from 1990 to 2001, in several regions of Ethiopia and, most recently, in Southern Sudan whose status as an independent state was recognised by the UN in 2011.[1] Somaliland has de facto seceded from Somalia, although it has yet to be recognised as its own country, even though it is possibly the only part of the country with a functioning government. Pierre Englebert explains that secessionist movements in Africa, however, are largely driven by economic and material motivation rather than ethno-political differences: most breakaway states have been areas with superior access to resources relative to the rest of the country. In their efforts to secede from the mainland, but constrained by international norms of sovereignty and recognition, breakaway states have repeatedly appealed to mercenaries to provide combat support and strengthen their armies. Katanga and Biafra are two examples of secessionist states that have hired mercenaries for this purpose.

After declaring independence from D. R. Congo in January 1961, Katanga's new 'President', Moise Tshombe, began recruiting foreign mercenaries to buff-up his army which was financially and politically supported by the Union Minière du Haut-Katanga (UMHK). Five hundred Belgian, French, Rhodesian and South African soldiers of fortune poured into the country. The Katangan mercenaries were referred to locally as '*les Affreux*', because of their lack of accountability and organisation. These men were led by former French colonel Bob Denard and Irish colonel 'Mad' Mike Hoare. Known as the *gendarmerie*, the mercenary force was a threat to the sovereignty of the newly independent D. R. Congo. When Moise Tshombe instructed the mercenaries to attack and kill UN Peacekeeping forces, the UN retaliated by sending in 5,000 men to drive all foreign combatants out of the Congo.[2] Journalist Tony Geraghty claims that the mercenaries "led by (Frenchman) Roger Faulques, had killed 1,000 U.N. troops when they attempted to end (the) secession" (Geraghty 2007: 41). Eventually, the mercenaries found themselves outnumbered by the UN and Tshombe agreed to lay down arms and reintegrate his province into D. R. Congo. The mercenary war further reinforced the norm against white mercenaries and brought about accusations of neo-colonialism, further strengthening Africa's resolve to combat this 'disease' eating up the continent.

Mercenaries also featured prominently in the Nigerian–Biafran Civil War. Lieutenant-Colonel Ojukwu's proclamation of independence of the resource-rich state of Biafra in 1967 provoked an outcry from the Nigerian government, and equally so from the foreign oil companies who had an interest in maintaining their influence in the country. The new state of Biafra was too wealthy to be left alone. Ill-equipped, out-manned and out-gunned by the Nigerian Army, Biafra resorted to hiring mercenaries to fight its war and turned to the French for support. Former legionnaire Roger Faulques, along with 100 French mercenaries, returned to Africa once more to participate in the secessionist war. The Biafran air force was created, trained, and operated by Swedish Count Gustav von Rosen who had "obscure delusions of changing Africa" (Venter 2006: 322). Concurrently, the Nigerian Army also recruited mercenaries, particularly for their own air force with Egyptian mercenaries employed to fly the larger bombers. Percy claims that because of the ideological principles of independence movements,

> the mercenaries stood in direct opposition to the great project of self-determination, not only because they often literally fought against it but also

because the idea of fighting for money was disturbing in an environment where the other players were deeply motivated by a belief in national liberation.

(Percy 2007: 189)

She concludes therefore that the "mercenaries in Biafra in the civil war between 1967 and 1970 were largely useless, despite all the attention paid to them by the international media".

Mercenary coups

Mercenaries are not only instrumental to select African leaders: empowered by their own military successes and taking advantage of the weak political institutions of post-colonial states, a handful of mercenaries have sought to overthrow sitting presidents and take over the country themselves. Although mercenaries Bob Denard, Mike Hoare and Simon Mann were each convicted for attempting to overthrow the governments of sovereign states, these men did not operate alone.

In 1978, Frenchman Bob Denard invaded the island of the Comoros, killing the incumbent president Ali Soilih and replacing him with former president Ahmed Abdallah. Although president Soilih had grabbed power himself in 1975 with the help of Bob Denard, his shift toward socialist economic policies earned him many enemies, including France, which terminated all aid and technical assistance to its former colony. The 1978 coup against Soilih was carried out by 43 trained mercenaries and was allegedly supported by the Rhodesian, South African and French governments.[3] President Abdallah made Bob Denard the head of the Presidential Guard and the Frenchman de facto ruled the country for eleven years, until one of his officers shot and killed the president in 1989. Denard was subsequently evacuated to South Africa, but attempted another coup d'état in 1995. This time, Denard was an embarrassment to the new post-apartheid South African regime who opposed mercenary coups, and they summarily notified the French authorities.[4] France subsequently sent their special forces to the Comoros, arrested Denard and sentenced him to four years in prison for "belonging to a gang who conspired to commit a crime" (BBC News 20/6/2006). During his trial, Bob Denard claimed that the French authorities had been aware of his coup, a fact that was recognised by the courts.

In a similar scenario, Irishman 'Mad' Mike Hoare was hired in 1978 to lead a coup d'état in the Seychelles on behalf of ex-president James Mancham. Mancham had himself been deposed a year before by his Prime Minister, France-Albert René. A successful businessman with strong ties to England and the United States, Mancham was able to muster financial and political support from his allies. The 53 mercenaries who were sent to the Seychelles, however, were disorganised and undisciplined, leading to an accidental shooting upon arrival at the airport. Panicking, the mercenaries ended up hijacking an Air India plane and flying it back to South Africa where the men were arrested. Mike

Hoare was found guilty of hijacking a plane and sentenced to ten years in prison. An International Commission appointed by the United Nations Security Council also concluded that the mercenaries had been supplied and encouraged by South African defence agencies.

A third, more recent mercenary coup attempt was carried out in 2004 by Simon Mann, a British businessman and former SAS officer. Approached in 2003 by an 'anonymous benefactor' through the intermediary of Lebanese businessman Eli Calil with an offer to carry out a coup in oil-rich Equatorial Guinea, Mann began to orchestrate an intricate plot involving arms dealers, politicians, covert agencies and a collection of international mercenaries. The mission was to overthrow the "tyrant" President Teodoro Obiang Nghema and replace him with Severo Moto, a former politician and Catholic priest from Equatorial Guinea. Mann's plan was a failure, however, as repeated indiscretions and unforeseen obstacles led to a loss of political support from the United States, Spain and Great Britain. On 7 March 2004, Simon Mann and 69 mainly South African mercenaries were arrested with a cargo plane full of weapons at Harare airport, after a tip-off from the South African intelligence agency. The ensuing scandal led to renewed hostility towards mercenaries in Africa.

Shifting anti-mercenary norms

Africa's experience with mercenaries is unique. No other continent has had such exposure to so much interference from private and foreign individuals. In Resolution 56/232, the UN General Assembly expressed its alarm and concern "at the danger that mercenaries constitute to peace and security in developing countries, in particular in Africa and in small states". Mercenaries have been hired by secessionist states, ousted leaders and independent entrepreneurs to illegally challenge the legitimacy of the government. Understandably, the continent harbours strict anti-mercenary norms and has been particularly pro-active in passing legislation condemning and punishing the recruiting, use, financing and training of mercenaries. In 1977, the Organisation of African Unity passed the Convention on the Elimination of Mercenarism in Africa. As of 2008, however, only 28 of the 53 members had ratified this Convention. This suggests that despite staunch rhetoric against the use of mercenaries, insecure leaders want to maintain the prerogative of hiring the services of foreign combatants when deemed necessary to the survival of their regime.

Despite the media attention that these three above-mentioned events attracted, "mercenary coups in Africa are a thing of the past" according to Greg Mills of the Brenthurst Foundation.[5] Indeed, in 1999, the Organisation of African Unity declared that it would no longer recognise leaders who came to power through a coup.[6] This new approach breaks with the sacred tradition of non-interference and a nearly blind respect of sovereignty, engraved in its Charter. The general consensus among academics and political analysts appears to be that white mercenaries and foreign coups have passed their prime. The end of the apartheid regime has shifted South Africa's foreign policies towards non-interference and outright condemnation of mercenarism, especially the recruiting of mercenaries inside the country. This is a dramatic change since South Africa had previously been heavily implicated in mercenary and covert activities throughout the continent. According to Annette Seegers, African leaders are still afraid of having their rule challenged by foreign mercenaries, which is why most countries, including South Africa, remain pro-active regarding any rumours of mercenary activities.[7] Finally, barring Simon Mann's foolhardy attempt to topple the government of a sovereign state, mercenary coups in recent history have been few and far between.

African leaders continue to rely on foreign combatants to buff up their armies in times of domestic crisis, however: former Colonel Gaddafi allegedly recruited heavily from the Tuareg tribes in Mali and Niger in his last attempts to resist the popular uprising and foreign attacks threatening his leadership in Libya.[8] On the other hand, mercenaries have gradually been replaced by a new breed: the more respectable, if controversial, private military and security contractors. As legal and established corporate institutions, foreign-owned private companies offer similar services as gangs of mercenaries, but under an umbrella of respectability and with a promise of professionalism and moderately superior accountability, not unlike the White Company discussed in Chapter 2.

Private military and security companies

Private military and security companies have been the preferred security option for many political leaders in Africa. Foreign owned, these companies claim to work only for legitimate governments and offer a selection of services that range from training and advising armies to gathering intelligence, providing operational support and participating in combat operations. Tim Spicer, the CEO of private military companies (now defunct) Sandline and Aegis, defines PMCs as "corporate bodies specialising in the provision of military skills to governments: training, planning, intelligence, risk assessment, operation support and technical skills" (Spicer in Percy 2007: 60). Peter Singer categorises the industry of private military and security services into three sectors: military provider firms, military consultant firms and military support firms, which he organises according to their proximity to the conflict. Military provider firms are distinguished from the other two categories because they "run active combat operations" whereas military consulting firms offer the same services as provider firms, with the exception that they "do not operate on the battlefield itself" (Singer 2007) and are therefore, in theory, not subject to the same degree of risks and exposure as the national army. The third category, the military support firms, provides alleged non-lethal services of logistic and technical nature, although this can also include intelligence gathering.

Private military companies have generally set their headquarters in 'Western' countries– the United Kingdom, the United States, South Africa and Israel playing host to the largest number of international companies. They are managed by retired officers from elite combat units, businessmen and former politicians.

The personnel that work for these companies are mostly recruited from former combat units, including but not limited to the British Special Air Service, the US Special Operations Force and the former South African Defence Forces. Dismissed soldiers from armies in ex-Yugoslavia and the former Soviet Union as well as soldiers from Latin American armies have recently been recruited into the units of these private military companies – possibly because they are cheaper than 'Western' soldiers. Because of the stigma attached to private military companies that associates them with mercenaries, and the media condemnation that companies like EO have faced, "the new freelance teams preferred to be known as private *security* companies rather than private *military* companies" (Geraghty 2007: 186). The actual agents that operate on the ground, however, tend to be the same veteran soldiers recycled from *ad hoc* mercenary operations or military coups: several of the mercenaries who were sentenced for taking part in the 2004 Equatorial Guinea plot are former Executive Outcome employees who now work in Iraq and Afghanistan for private military and security companies in an 'advisory' capacity.[9]

Not unlike previous mercenary units, private military and security companies have enjoyed limited support from the governments of the countries in which they are established: EO and Sandline had to receive government licenses to manage their contracts and operations from their home state. The 1997 Regulation of Foreign Military Assistance Bill allegedly aimed at closing down EO did not stop the South African government from granting a licence to the company to continue its operations, according to its CEO Eeben Barlow. Since the purpose of the bill was to "prevent South African companies and citizens from rendering military or military-related services abroad without the Government's authority" (Malan and Cilliers 1997), by granting a licence to EO, the South African government was explicitly showing its cooperation with EO's activities. Tim Spicer also emphasised his company's close relationship with the British government: "we had been in regular contact with the Foreign Office and the US State Department and had always kept them fully informed" (Spicer 2000). Abdel Fatau Musah, a conflict prevention advisor at ECOWAS, further claimed that "although the British government has consistently denied working with Sandline International (…) there can be little doubt that unofficial links continue to dominate the 'old boy network' relationship between serving officers, diplomats and retired diplomats" (Musah and Fayemi 2000: 19).

Since EO's first contract with Angola in 1993, private military and security companies have met with substantial financial and political success. Private companies were used extensively by the various governments of Sierra Leone from 1994 to 1998. Originally hired to push back rebel insurgencies, private military companies have diversified their operations: they are now employed to defend important sites of oil and mineral exploitation, to protect the royal and presidential families of select countries – Vinnell is reported to train and advise the Saudi National Guard "which functions like a Praetorian guard to the regime" whereas "O'Gara protects the royal family" (Singer 2007) – and to train and rebuild the armies. The private security company DynCorp has been contracted by the United States to create an army of 2,000 soldiers in post-war Liberia.[10]

Since the early 1990s, the emergence of private military and security companies as a serious actor in strategic warfare has raised issues of military identity, accountability, reliability and sovereignty in weak African states. The privatisation of military services has also challenged the monopoly over the legitimate use of violence that states have traditionally claimed as their exclusive right. Despite legal conventions and norms prohibiting the use of foreigners for military operations, PMCs have prospered on a continent where political survival and perpetual conflict undermine international law and public opinion.

A case study of hybridisation in Angola

Angola was the first African country to experiment with hybridisation. After more than twenty years of civil war and under the imminent threat of being overrun, the national armed forces Forças Armadas Angolanas (FAA) along with acting President José Eduardo dos Santos jointly decided to hire the services of the private military company Executive Outcomes (EO). Run by a South African intelligence agent, Eeben Barlow, and registered in the UK, the company offered combat services and military advising to their clients. Between 1992 and 1996, EO actively supported the FAA against the rebel insurgency. The four-year partnership received intense media and academic scrutiny. Consequently Angola offers a unique opportunity to study the interactions between the national armed forces of Angola and the civilian combatants working for the private military company.

Background to the conflict

Angola officially became a Portuguese province in 1951, after more than 500 years of close interactions with the Kingdom of Portugal, including a stint of 400 years as a Portuguese colony. As a colony, Angola experienced the development of the mining and oil industries by the Portuguese, the forced Christianisation of the population, the installation of a remunerated forced labour system and a complete disenfranchisement of the Angolan people who were excluded from political life, from the armed forces and from the economic development of their country. This inevitably led to a popular uprising against the Portuguese occupation, starting with the formation of opposition parties in the 1950s and 1960s: the Movimento Popular de Libertação de Angola (MPLA), the Partido Comunista Angolano (PCA) (which eventually merged with the MPLA), the Frente Nacional de Libertação de Angola (FNLA) and the União Nacional para a Independência Total de Angola (UNITA) led by Jonas Savimbi were organised along ethnic lines that had been exacerbated through the Portuguese's preferential system of division of labour.

The Portuguese Army responded to this challenge with violence and repression. The ensuing war of independence lasted until 1975, during which time the three aforementioned parties fought each other and the Portuguese with equal fervour, using guerrilla tactics. Independence was finally achieved after the Portuguese' right-wing government, the Estado Novo, was overthrown in a military coup, later referred to as the Carnation Revolution. This brought about a liberal regime in Portugal that granted independence to the state's three colonies: East Timor, Mozambique and Angola.

The Angolan civil war took off immediately after independence, with both the MPLA and UNITA arbitrarily claiming sovereignty over the territory. The MPLA, formed by an "educated left-wing, urban elite concentrated in Luanda" (Cilliers and Mason 1999: 143), controlled most of Luanda, the capital city, and adopted a Marxist ideology bringing in political, military and financial support from the USSR and Cuba. They installed Agostinho Neto as the first President of Angola; he was succeeded after his death in 1979 by José Eduardo dos Santos. Dos Santos has been the acting president ever since. UNITA, with its base among the rural, largely peasant territory of Angola, responded by setting up a rival government along with the FNLA and enlisted the help of South Africa and the United States as a counter-force to the Marxist-Communist 'threat'. The South African Defence Force "co-operated for the entire period between 1975 and 1989 with Jonas Savimbi's UNITA rebels, who were engaged in a civil war against Angola's self-proclaimed Marxist MPLA government" (Mills and Williams 2006: 169).

South Africa's commitment to UNITA was a reaction both to the MPLA's support of SWAPO (South West Africa People's Organization) insurgents calling for the independence of Namibia – a South African territory at the time – and to the perceived threat of a Marxist Angola on the internal politics of the South African apartheid system. Likewise, Cuba maintained both an ideological and a financial interest in Africa, in particular in Angola where it had a wide margin for interference. Between 1975 and 1988, the Angolan war was a "Cold War proxy conflict (...) reflect(ing) the classic parameters of 20th century conflict" (Mills *et al.* 2003: 7) with both the 'recognised' government and the rebels receiving weapons, financial support and military support from various states. This inevitably led to a vicious circle of unadulterated violence provoked by foreign governments who used the Angolan conflict as a forum to fight out their own ideological differences. The civil war and foreign interest were further fuelled by Angola's vast resources: the country is the second largest oil producer in Africa, after Nigeria, and the fourth largest diamond producer in the world. It also has an enormous coast line of 1,650 km. On the other hand, Angola has an inadequate supply of drinking water and only a small area of arable land. Access to mining and oil concessions has furthermore drawn competing countries and companies into Angola's civil conflict.

In 1983, the UN Security Council demanded South Africa's withdrawal from Angolan territory, which activated the signing of the Lusaka Accords in which South Africa agreed to remove its troops on the condition that the MPLA cease all support for SWAPO. This initiative failed and the South African Defence Force was sent back to Angola in 1985 to support UNITA's forces against the joint MPLA/Cuban alliance: "by the mid-1980s, with substantial South African backing, UNITA's operations had grown to the point where they posed a considerable threat to the security of the MPLA" (Mills and Williams 2006: 172). This culminated with the MPLA's defeat at Cuito Carnavale in 1988 at the hands of a numerically inferior SADF/UNITA coalition. The cost of this military campaign, however, compounded by international sanctions against the apartheid regime, was beginning to take its toll on the South African Defence Force and the political will of the government and its citizens.

Financial backing on the part of the Soviet Union and Cuba was also beginning to fail and in 1988, "a cash-strapped Soviet Union, beset with its own growing internal problems, told the government of President dos Santos that is was no longer prepared to provide weapons and assistance" (Barlow 2007: 95). Consequently, in December 1988, South Africa, Angola and Cuba signed the Brazzaville Protocol which brought about the implementation of UN Security Resolution 435 for a ceasefire and the peaceful transition towards the independence of Namibia. This removed South Africa's direct interest in supporting UNITA, although continued interest in the resource management of the country allegedly remained an important aspect of South Africa's foreign policy.

Elections and a return to war

Dos Santos and Savimbi agreed to meet in January 1989 and brokered a peace deal which was immediately rescinded over "a disagreement about what their oral undertakings had been about and over Savimbi's future role in the country" (ibid.: 98). After the collapse of the Soviet Union, the MPLA's rejection of its Marxist ideology and a series of peace talks, the warring parties reached an agreement known as the Bicesse Accords, which 'guaranteed' a free and fair election. The first and so far only 'free and fair' election in Angola took place in September 1992. The international community generally agreed that the process was indeed democratic: 91 per cent of registered voters, i.e. 4.4 million people, turned out to vote. Dos Santos received 49.6 per cent of the vote against Savimbi, who retained just 40.7 per cent of the electorate. As both candidates received less than a majority, a second round of voting was required, although this never occurred as Jonas Savimbi and UNITA rejected the results and returned to war.

According to former South African diplomat Sean Cleary, Savimbi was under the impression that the elections had been fraudulent because he controlled 75 per cent of the country and therefore should have easily won a majority of votes.[11] Furthermore, he states that the MPLA, threatened by Savimbi's position of relative power, "attacked and destroyed all UNITA's residences and party offices in Luanda, leading to the death of many and the capture of almost all its military and civilian cadres in the capital" (Cleary in Cilliers 1999: 153) between 31 October and 2 November 1992. This is corroborated by Simon Mann in *Cry Havoc*: "as the news of the election result came through, UNITA supporters and staff were hunted down and slaughtered. Savimbi himself narrowly escaped".

The MPLA and Simon Mann argue that the massacres were planned by UNITA to give them an excuse to return to war. Both sides mistrusted the other and reacted to any provocation with unrestrained violence. The UN by this time had already lost credibility, having maintained an insultingly small force in Angola since 1989, including "350 unarmed military observers, 130 unarmed police observers, and 100 electoral observers" (ibid.: 145).

Violence and insecurity

By the time the private military company EO was contracted in Angola in 1993, the MPLA was desperate: the FAA had been severely weakened as it had proceeded to disarm and reduce its numbers in accordance with the conditions of the Bicesse Accords, and was unable to contain the military threat when UNITA took up arms after the 1992 elections. In mid-1993, UNITA controlled 75 per cent of the country and the MPLA, despite having won the first round of elections, was losing the war.[12]

In 1995, Human Rights Watch published a report on the human rights development of Angola: it estimated that more than 100,000 people had died, including

> 250 child deaths reported each day in the besieged government-held city of Malanje alone. In September 1994, the U.N. Secretary-General reported that there had been a 10% increase in the number of people severely affected by the war since February 1994, and that nearly 3.7 million Angolans, mostly displaced and other victims of conflict, were in need of emergency supplies, including essential medicines, vaccines and food aid.
>
> (HRW World Report 1994)

UNITA was accused of indiscriminate shelling of cities, using civilians as a weapon of war and recruiting child soldiers. These tactics were allegedly equally used by the government's MPLA forces: "the ethnic cleansing of Ovimbundi and Bakongo citizens, revenge killings by both sides of those suspected of supporting the other in cities which changed hands; as well as from landmines, starvation and disease" (Cleary in Cilliers 1999: 146). UNITA was known to lay sieges to cities and towns, causing widespread starvation. It was also infamous for attacking humanitarian relief operations.

Both sides admitted to torturing prisoners. Kidnappings, especially of foreigners, were widespread. Mine warfare was rampant with

> thousands of new mines being laid by the government and UNITA to obstruct roads and bridges, to encircle besieged towns with mine belts up to three kilometres wide and to despoil agricultural lands. There were an estimated nine to fifteen million mines laid throughout the country. The U.N. estimated that the number of amputees as a result of mines injury will reach 70,000 in 1994.
>
> (HRW World Report 1994)

UNITA financed its military campaign by occupying mining areas and leasing concessions to foreign industrialists or mining and selling the diamonds themselves. This strategy also deprived the government of its main source of income.

Not only was Angola weakened by a quarter century of civil war, it also suffered from intrinsic weaknesses characteristic of the colonial system which further increased its vulnerability to security threats. The United States Department of State highlights the high level of corruption in the country. Angola placed 158 out of 180 countries on the Corruption Perception Index in 2008: "corruption is rife at all levels of government – mainly owing to the meagre salaries paid to low-level officials" (Grobbelaar 2003: 51). Police, government employees, immigration and customs officials are known to frequently and regularly ask for bribes and accompany their abuse of authority with violence: the United States travel website warned that "officials are sometimes undisciplined; however, their authority should not be challenged". The Angolan government is unable or unwilling to provide security throughout the country, leaving both foreigners and citizens at the mercy of gangs, rebels and miscreants:

> several attacks against expatriates in Cabinda resulting in rape, robbery, and murder were registered. Those responsible declared their intention to continue attacks against expatriates. Occasional attacks against police and Angolan Armed Forces convoys and outposts in Cabinda also continue to be reported.

Similarly, the diamond interests further contributed to weakening the state as "capture and occupation of these areas for personal benefit were important goals" for FAA generals and officers as much as for the rebels (Cleary in Mills 1995: 163).

Angola's criminal justice system requires institutional reforms: "Angola has only 656 registered lawyers (…). Luanda has five judges, each of whom deal with an estimated 900 cases per year (…). External awaiting-trial periods of between two and three years are common for prisoners" (Grobbelaar 2003: 50). Furthermore, the government's failure to guarantee the salaries of its police and armed forces have turned these agents of safety and security into a threatening force towards their citizens: According to Neuma Grobbelaar,

> the police have also allegedly been involved in widespread criminal activity. The latter is largely attributable to failure by the government to pay the salaries of security forces, which leads to their harassing, extorting and abusing civilians to obtain supplies, with tacit support from the government.

Consequently, citizens have very little faith in their undemocratic government, which is itself unable to control the armed forces and indeed has set them against the very people they should be protecting.

Angola displayed all the symptoms of a failing post-colonial state: the government had no legitimacy and very little authority. The armed forces were weak and undisciplined, unable to claim a monopoly over the legitimate use of violence. Finally, the population were the ultimate victims of the war, but remained impotent agents caught between the rebels and their own government.

Outsourcing security

It was in this environment of violence and chronic structural failure that President dos Santos and General Luis Façeira, the Angolan supreme commander of the army's ground forces, contacted Eeben Barlow in April 1993, offering the South African a contract for his company to train the MPLA's army and defeat UNITA once and for all. EO had already started to build a reputation for itself, after a successful contract with Heritage Oil & Gas for which it had provided security and recovered valuable material from Soyo, an area in northern Angola held and controlled by the UNITA rebels. Heritage Oil & Gas was heavily involved in oil exploration and drilling of the Angolan coast, and its CEO, a British businessman by the name of Tony Buckingham, therefore had an interest in the political development of the country.

Tony Buckingham and his number two, Simon Mann,[13] were present at the meeting between Barlow, President dos Santos and General Façeira. The two oil moguls allegedly ended up supplying most of the financing for EO's contract in Angola: "Tony Buckingham would guarantee all funds and ensure they were paid into my company account as stipulated. The finance would be provided by Sonangol, Heritage Oil and Gas, and Ranger Oil" (Barlow 2007: 138). The one year contract was eventually extended to four years, but there appears to be contradictions regarding who paid EO's fee of approximately US$60 million: the MPLA, the oil companies, or a joint-venture between the two? Allegedly the contract was sponsored half-half by Tony Buckingham and the MPLA.[14] There have also been allegations that the contract was partially paid with mining concessions: "in further payment for EO's services, substantial concessions were granted to Branch Energy, Buckingham's company" (Cleary in Cilliers 1999: 163). On the other hand, according to Simon Mann, EO's first contract with Tony Buckingham's Heritage in Soyo was paid for by the Angolan government.

The decision to outsource security to a foreign private military company was a deliberate policy adopted jointly by the head of state, President dos Santos, and the head of the armed forces, General Luis Façeira. According to Barlow, the decision makers in Angola recognised that they

> were in desperate need of help. We have made many requests to the United Nations which they have ignored. No one there wants to help us. When UNITA returned to war after they lost the elections they caught us completely off-guard. We had demobilised our units, which allowed UNITA a free hand to occupy large areas of Angola.
>
> (Barlow 2008: 138)

Their perceived abandonment by the international community and the inability of the army to defeat the rebels drove dos Santos and Façeira to contract EO with the belief that the mercenary company would be able to supplement the national armed forces by bringing in a thorough training programme, clear leadership and logistical efficiency. EO would thus act as a force-multiplier and change the course of the war in the government's favour. General Antonio Façeira of the country's Special Forces commented that "this is probably our last chance of saving our country" (ibid.: 135), showing that the armed forces recognised their inability to control the situation and their desperate need for effective military support.

This narrative, however, is contested by Cleary, who argues that the FAA was a "highly capacitated military force with sophisticated equipment, good order of battle" which had benefits for Soviet and Cuban advisory groups.[15] Cleary suggests that Angola suffered a massive political failure in which rampant mistrust between the two parties, UNITA and the MPLA, led to a self-sustaining prophesy of incapacitated peace negotiations. He further explained that when UNITA took over Soyo, a major oil producing city, the government and oil companies turned to EO as a quick, easy and effective solution to retake the town. Although the FAA could have handled UNITA, which was already marginalised and had relatively little access to resources beyond some diamond fields, the senior ranks of the Angolan armed forces benefited far more by outsourcing security and encouraging a continuation of the war than by suing for a sub-optimal peace: In his autobiography, Barlow explains that

> De Matos will not accept UNITA's army integration into the FAA, despite what has been agreed on the political level. Full integration is not possibly as De Matos will then not be able to count on the loyalty of the FAA in the event of a return to war. De Matos benefits from EO's presence in Angola, it reinforces his strength as they are answerable to him. He has factored them into his planning for the reconstitution and training of FAA.

In fact, the diamond interest fuelled the war to a large part as the above-mentioned military officers gained access to mining concessions which are still owned by army veterans today. The army was also able to participate in advantageous weapons purchases through their cooperation with EO. Undoubtedly a strong component of hiring EO on the part of the military officers was fuelled by the possibilities for career advancements in times of war, access to military technology and high-tech weapons, expert training and diamonds. President dos Santos would also have appreciated that the cost of hiring a private military company was cheaper and faster than investing in military training, purchases, pensions and health care, and does not require immediate access to cash. Greg Mills further explained that in 1992, the FAA might have had the equipment, but they certainly did not have the training or the money; EO gave them an advantage that was able to swing the balance in their favour.[16]

EO's military campaign

During the four years that EO spent in Angola, it developed a comprehensive training programme for the FAA, the Angolan Army. This included intelligence, counterinsurgency operations, military discipline, small unit tactics, administrative, logistic and maintenance training, leadership, offensive and defensive exercises. Although it appears that the initial contract was for the purpose of training the Angolan Army, a joint EO/FAA plan "aimed at defeating UNITA on the battlefield" (Barlow 2008: 189) saw the mercenaries lead the newly trained Angolan soldiers in successful campaigns against UNITA as they proceeded to retake various parts of Angola, starting with the oil-producing and diamond mining areas, Saurimo and Cafunfu.

The presence of EO and their rigorous training programme gave the FAA a much needed boost of confidence, and the Angolan forces resumed the war with enthusiasm. Furthermore, EO provided their employees with constant air support. This was important because it motivated the soldiers to take the risks necessary to carry out their mission: nobody went into the bush without asking who was flying the helicopters or planes in case of an emergency evacuation.[17] The "fixed wing and helicopter" also gave the government forces an edge as they flew "countless sorties, often braving intense anti-aircraft fire to relieve pressure on the FAA" (Barlow 2008: 88). The Angolan armed forces were to provide all weapons and equipment needed to conduct the training. This included "a full range of weapons, vehicles, armoured personnel carriers, helicopters and ground-attack aircrafts" (ibid.: 136) that were generally bought from third countries. One EO pilot suggested that the Angolan Army had all the modern weaponry, including helicopters and fighting planes, that they needed, but lacked the organisational ability, logistical know-how and leadership to utilise their material effectively. South African veteran, Colonel Velthuizen, pointed out that it was the low level of maintenance and lack of technicians that rendered these weapons useless, which partially explains why the FAA required foreign support to fight their war in the first place.[18]

By November 1994,

> UNITA had been convincingly defeated by the FAA over a wide front on the battlefield. They had lost the strategic diamond mining town of Cafunfu and were coming under increasing pressure to accept a ceasefire. The MPLA was finally able to negotiate from a position of strength.
>
> (Barlow 2008: 290)

UNITA and the MPLA signed the Lusaka Protocol on 20 November 1994 making provisions for "a complete cease fire, the integration of UNITA's general officers and other ranks into the FAA, UNITA's demilitarisation, the repatriation of mercenaries" (ibid.). After overwhelming pressure from the international community and in particular the United States, EO left Angola in January 1996, but not without warning the government that the situation was still too precarious to leave. Indeed, in June 1997, UNITA resumed its offensive campaign, attacking a village in Luanda Norte and killing all its inhabitants. The civil war raged on, virtually ignored by the international community, until the death of Jonas Savimbi, reportedly killed by the MPLA, although this has been contested orally by some mercenaries and South Africans working in Angola who claimed that "Israel took him out".[19]

EO claims that its intervention forced Savimbi and UNITA to the negotiations table, a statement which is supported by Herbert Howe. Between 1993 and 1996, the time when EO left the country after its contract had ended, the military company had "trained a demoralised and defeated army and turned the tide of the 20-year-long Angolan war" (Barlow 2008: 299). Cleary, on the other hand, argues that EO gave the FAA a military advantage beyond its capabilities, and therefore "contributed to the prolongation of the war".[20]

Analysis of legitimacy

EO claims that it only worked for legitimate governments – but in weak African states, this was a big risk, as the government of the day could change very quickly and democratic elections were few and far between. From its independence in 1975 to the elections in 1992, Angola was represented by various factions, each supported and recognised by different communities in the international system: UNITA was backed by the United States and South Africa, whereas the MPLA could count on the support of Cuba and the Soviet Union. In 1976, the Organisation of African Unity (OAU) formally recognised the MPLA as the legal sovereign of the territory and the official representation of Angola abroad, after the party overpowered several UNITA strongholds. The two parties eventually came to an accord and agreed to a ceasefire and elections in 1992. These UN-monitored elections, however, led to an impasse as neither candidate received a majority and consequently a second round of elections with a runoff between the two principle candidates was required by Angolan law. The second round of elections never took place because the civil war resumed shortly after. The international community subsequently condemned UNITA for its failure to honour the election results, and promptly recognised the MPLA as the legitimate government.

The MPLA therefore derived its right to rule from international legal sovereignty. The international community's recognition sufficed for it to claim development aid and support and the right to organise and distribute the resources and manage the population within its own territory. The MPLA could subsequently set its own foreign policy, interact with foreign investors, distribute mining concessions and fix levels of taxation in all 'legitimacy'. Within its own territory, however, the reality was quite different: the MPLA was unable to effectively control the diamond mines which were taken over and exploited by UNITA. Indeed, by 1993, UNITA allegedly controlled some three-quarters of Angolan territory, although arguably most of this land was unpopulated waste land which explains why the Angolan economy continued its strong growth during the civil war: the MPLA was able to control the capital city, Luanda, and the oil fields from which it derived its main revenue. It also held the monopoly on the licensing and taxation for the trading of commodities in and out of the country.

Nonetheless, the MPLA was never able to qualify as the legitimate government in terms of Weberian sovereignty. The party did not provide its population with security, nor was it capable of monopolising or even controlling the level of legitimate violence within its territory: the MPLA therefore did not "successfully claim the *monopoly of the legitimate use of physical force* within a given territory" (Weber 2004) beyond its stronghold in Luanda. Instead of encouraging a social contract between the government and its citizens, the MPLA criminalised all forms of opposition, and then enacted its policy with violent enthusiasm against UNITA and any potential supporters, at times attacking its own population. It failed to provide its citizens outside of Luanda with most public goods. Furthermore, despite its relative capacity, the Angolan Army was incapable of controlling its borders, enabling illicit support for UNITA and illegal mining and trading to occur within its territory. This encouraged UNITA to continue fighting the MPLA as it was able to keep its supply lines open, facilitating its military efforts by exchanging weapons for illegal diamonds. Professor Seegers argues that capacity depends on one's relations with its neighbours,[21] and UNITA was able to support its campaign with the help of its foreign allies, in particular Mobutu's regime in Zaire which equipped the UNITA forces and happily traded with them.

In countries that do not necessarily have a history of democratic regimes and good civil–military relations, governments have very few avenues to claim legitimacy: they can have legitimacy stamped on them by the international community but this does not translate into effective sovereignty in the sense that they often encounter internal dissent from civilians or the military. Alternatively, the government can seize control by force and claim Weberian sovereignty by their capacity to monopolise the legitimate use of violence and provide security or alternatively threaten to repress its citizens. This was the case in Liberia, although the international community's failure to recognise Charles Taylor's regime did undermine his foreign policy. The government can also claim legitimacy by holding elections and wining a majority of votes. By this last strategy, it effectively lays the groundwork for a binding legal relationship between itself and the people it claims to represent. These elections are not necessarily democratic, but they are symbolic and have come to represent a compromise between the international community, the government and the population. Indeed, most peacekeeping missions are principally aimed at promoting and facilitating elections that subsequently confirm the government in power and enable commercial and political relations to resume. The relationship between a government and its people is therefore based on political control, not a social contract as idealised by the West.

Because of the failure to hold a second round of elections, the MPLA never cemented its right to represent Angola through a democratic process. Legally speaking, the MPLA did not receive its mandate from the people as it could neither claim to represent the majority of the population nor to have respected the constitution's electoral requirements. The MPLA's armed forces were also unable to provide security to the population or to monopolise the legitimate use of force, as evidenced by UNITA's ongoing military threat against both the citizens and the state. In fact, the MPLA derived its legitimacy entirely from the international community, despite its obvious failures to carry out the mandate expected from sovereign entities. President dos Santos has ruled Angola from 1979 and is still in power in 2014. This suggests a breakdown in the democratic process of the country, and undermines the legitimacy of the MPLA, despite its alleged victory in subsequent elections. Eeben Barlow's and Simon Mann's claims that they worked for a legitimate government can therefore be contested in this environment of contrasting and overlapping claims to sovereignty, and require a more developed understanding of the political context in which the private military company is operating.

EO was working for dos Santos and the MPLA, which were the internationally recognised authority in Angola, and thus the only legal government in the eyes of the United Nations. Consequently Mann and Barlow can unquestionably claim that they were on the side of international law and international legal sovereignty, in this case, despite the general norm at the time that stigmatised private military companies. Since the MPLA was unable to fulfil the mandates generally expected from a sovereign government, however, its claim to internal sovereignty through the use of a foreign military agent suggests an unlawful attempt to grab power beyond its actual capabilities. EO and its administrators may be in their legal right, but the impacts they have on the legitimacy of the state and its relationship and responsibilities towards its own citizens are politically questionable, especially in the long run.

PMC-military relations

The FAA was reputedly a highly capacitated, equipped and capable military by African standards. Annette Seegers explained that the Angolan Army is "excellent, organised, and probably one of two armies which could take on the South African Army". General Façeira's decision to collaborate with President dos Santos and hire EO to supplement the armed forces is an unusual example of civil–military cooperation. It also ensured the success of EO's mission without overtly threatening the identity or the authority of the Angolan armed forces. EO had a distinct advantage, as most of its personnel had previous experience fighting in Angola from the time when South Africa had supported UNITA. This gave the men personal knowledge of the terrain as well as intelligence into the tactics, strengths and weaknesses of their new enemy, but it also fuelled confusion and anger among the FAA who had previously fought against the SADF soldiers: "their initial fear and distrust of my men was based largely on the stories of their fathers and older brothers who had once been pitted against the South Africans on the battlefield" (Barlow 2008: 154). It should be noted, however, that this was a unique circumstance and EO's experience in one

African country does not make their team more or less capable of operating in another context, such as Sierra Leone, where the terrain and socio-military infra-structure is so very different.[22]

Dealing with a foreign private military company can be a threatening and emasculating experience for any national army. Reactions range from relief "thank God someone here is competent"[23] to embarrassment and outright rebel-lion. According to Jane's defence correspondent Helmoed-Romer Heitman, "the guys who are actually good will have a problem and question the command structure". By bringing in a PMC, the government sends a message that its army is severely limited, and this both undermines and disrupts the morale of the armed forces, leading to dissent. The tensions between the foreign forces and the national army are further exacerbated by the perceived nature of mercenarism: "mercenaries are boastful bums with little discipline".[24] Eeben Barlow acknow-ledged this problem and explains in his autobiography that it

> pained many of them (FAA soldiers) to see large numbers of ex-SADF sol-diers in their midst. Although General De Matos had instructed them that our men should be viewed as friends and allies, it was difficult for many FAA soldiers to accept.

As a result, some soldiers "refused to issue weapons, ammunition and other equipment to my men (and) there was even talk of some of our men rebelling at the way they were being treated" (Barlow 2008: 152). Both the FAA and EO personnel faced great challenges in their interactions, despite the initiative being spearheaded by the senior officers of the Angolan Army. Nonetheless, EO suc-cessfully completed the training of 16-Brigade and led them into battle, most notably in Saurimo and Cafunfu. The experience of battle contributed to forging a shared experience of suffering and excitement, improving the relations between the soldiers and the foreign combatants. Taking part in offensive operations was beyond the contract of EO, but nonetheless was arguably instrumental to the success of their mission: "our men formed one of the brigade's combat teams. When we initially recruited them, they were there purely to train the FAA. Now they're gearing up for war" (Barlow 2008: 227).

Another problem that is characteristic of PMC–military relations is the ranking system. PMC personnel are outside of a national military structure. Con-sequently, even when they are 'promoted' to various ranks, this lacks the official sanctions of government recognition and undermines the credibility and the merit of the soldier. The national armed forces are understandably sceptical of foreign contractors who emerge with glossy ranks and give them orders without any formalised structure. A further concern which cannot be overlooked is that the private military industry also keeps official troops out of the promotion line by taking over the role of the armed forces. This is a straightforward threat to the livelihoods of these soldiers who, very often in these countries, are barely paid their salary and have little margin for promotions, not to mention a pay-rise. To overcome trust issues with senior Angolan officers, Eeben Barlow attributed equivalent ranks to his own senior personnel, regularly briefed the generals on the training programmes and operations and held joint-planning sessions. These briefings "played a major role in establishing a lasting foundation of trust with the General Staff" (Barlow 2008: 180). This maintained a relationship based on respect that did not undermine the command structures of the Angolan Army. The FAA and EO shared information and cooperated closely, therefore improving their access to tactical intelligence and drafting military plans that satisfied both parties.

Civil–military relations

EO also focused on building trust among the civilian populations. This was necessary to maintain popular support for their operations and encourage informants to cooperate with the FAA and turn away from UNITA. EO aimed to set itself as a source of security for the local population wherever they were stationed. They also gave instructions to their engineers to supply purified water and tasked their medics to provide medical care to the communities: "the EO medics tried to assist the injured and malnourished *povo* wherever they could, despite being short of supplies" (Barlow 2008: 245). The result was that "people started waving at us when we drove past, happy to have us around when they developed any medical problems" (ibid.: 155). Although the strategy is necessary for a 'hearts and minds' campaign in which the counterinsurgency requires support from the local population, this does not appear to be the objective in Barlow's short-term contract in Angola. Rather, the impact that EO had on the local population was arguably opportunistic and short-lived, and appears to be an initiative led by the teams on the ground out of a spirit of compassion and to encourage the people to cooperate with the troops: "trying to relieve the misery of the locals in Angola was not to our financial advantage – nor was it part of our contract" (ibid.: 245).

Tim Butcher, defence correspondent for the *Telegraph*, argued that the mercenaries were actually feared by the local population because of their unaccountability, the fact that they "didn't give a shit", and because they were outsiders.[25] Locals were weary of their presence, and while the PMCs may have contributed somewhat to their immediate surroundings, they did the absolute minimum. According to Butcher, the hearts and minds campaign is "an absolute myth" and corporate social responsibility for a private military company is the height of irony: "I never heard anybody say 'bring them back' ".[26] Barlow's brief description of his company's contribution to the local community seems to confirm his lack of commitment to this particular enterprise. Greg Mills paints a different picture, however, stating that the people just wanted the conflict to end, and did not care whether this was achieved through the use of mercenaries or foreign soldiers.[27]

Conclusion: a zero-sum equation

The Angolan civil war lasted 27 years and officially came to an end on 22 February 2002 with the assassination of Jonas Savimbi. During nearly three decades of conflict, an estimated half a million people were killed, "more than four million (33%) of Angola's population are today internally displaced, and 450,000 are refugees (…). Nine million of its 13 million people live on less than a dollar a day" (Grobbelaar 2003: 15). The damage to the infrastructure and political and economic institutions is immeasurable. Today's estimated debt burden of US$11 billion largely reflects the MPLA's war effort since 1990.

For Angola, the decision to hire a private military company does not appear to have had a significant impact on the long-term security of the state. In terms of government credibility, dos Santos' initiative of hiring a private military company had little or no impact on his regime or on his relationship with either the armed forces or his citizens. Despite EO's short-term military successes, the civil war carried on until 2002 with UNITA continuing its assaults against the MPLA. EO's training programme, however, arguably made the armed forces more effective, as the MPLA subsequently launched a campaign against the rebels: a "ruthless scorched-earth policy towards the civilian population in areas traditionally supportive of UNITA, an effective UN-supported international sanctions campaign that froze its financial assets, isolated it politically, and embarrassed governments traditionally supportive of UNITA" (ibid.: 1) contributed to toning down the conflict, if not ending the war entirely. The end of Mobutu's regime in 1997 also undermined foreign support for the rebels. Eventually, the "stick-and-carrot approach followed by the Angolan government led to large-scale desertions by UNITA soldiers after the introduction of an amnesty in 2001 and, later, financial assistance 'for all those who abandon unjust war and opt for democracy'" (ibid.). Cleary claimed that "EO unquestionably helped the FAA to achieve its military objective": while EO may have contributed to prolonging the war, it cemented the FAA as the dominant military force in the country. This effectively reinstated the Weberian sovereignty of President dos Santos in Angola.

EO encountered all the expected obstacles of a foreign private military company operating in a developing state with weak political and military institutions. The company's success in overcoming these problems is largely due to its excellent managerial initiatives that focused on cooperation and communication between the army and the private combatants. By including the leadership of the FAA in all stages of planning, EO set itself in a support role that neither threatened nor undermined the authority of the national armed forces. The combative role and additional air support that EO supplied demonstrated the company's willingness to lead by example and won the trust of the Angolan soldiers. This episode suggests that private military companies can, at the very least, *not* adversely affect the civil–military relations of country in which they are operating. This depends, however, on the management of the company, its ability to win credibility and lead respectfully, and of course, on its mandate and *modus operandi*.

Finally, mercenaries and private military companies have found that Africa offers many lucrative opportunities, both because of the huge availability of resources, and because of its weak institutions and conflict-prone environment. Errors and abuses, mercenary coups and secessionist activities have contributed to poisoning the reputation of (particularly white) mercenaries and stigmatising foreign combatants as opportunistic and unscrupulous neo-colonialists. Nonetheless, nowhere have these foreigners encountered more freedom of action than on the African continent, where legislative efforts and international interest have largely veered away from implementing any anti-mercenary laws whenever there is a perceived advantage to hiring the services of mercenaries or private military companies. American, French and British PMCs have generally received the political support of their home state in their African contracts and only South African mercenaries have really been targeted by international and domestic condemnation. Despite international complacency, Africa's mixed experiences with mercenaries and private military companies show that these foreign combatants do not appear to make a significant difference on the ground.

The Angolan case study on hybridisation illustrates the experiences of a developing country with weak political, legal and military institutions. The next chapter investigates a new model of hybridisation taking place in a collapsed state where the expeditionary force and private military companies work together for a state that is remote from the theatre of operations.

Notes

1　The ethnic conflict waging in South Sudan in 2014 testifies to the serious and dire results of international state 'making' in Africa.

2　The voting of resolution 161 in the Security Council brought UN peacekeepers to the area with the mandate of taking any necessary means, including violence, to drive all foreign combatants out of the Congo.

3　For more, see Paul L. Moorcraft and Peter McLaughlin 2008 and Pieter Wolvaardt, Tom Wheeler and Werner Scholtz 2010.

4　Interview with Tom Wheeler, Johannesburg, April 2011.

5　Interview with Greg Mills, Johannesburg, April 2011.

6　As recently as 2013, however, the African Union failed to take any action when rebel leader Michael Djotodia overthrew sitting president François Bozizé.

7　Interview with Professor Annette Seegers, Cape Town, May 2011.

8　Interview with a Former South African Mercenary, Johannesburg, April 2011.

9　These allegations were repeated in several interviews with different contractors.

10　Interview with Christian Bock, Former Employee of Dyncorp, Cape Town, April 2011.

11　Interview with Former Diplomat Sean Cleary, Cape Town, April 2011.

12　This figure was publically claimed by Jonas Savimbi, although as Barlow points out, "the mere absence of the FAA in certain areas did not imply that UNITA dominated them" (Barlow 2007: 184).

13　Simon Mann started off as Tony Buckingham's "office manager (...) and wannabe oil tycoon" in April 1992 (Mann 2011).

14　Interview with Niel Steyl, Former Executive Outcomes Pilot, Pretoria, April 2011.

15　Interview with Former Diplomat Sean Cleary, Cape Town, April 2011.

16　Interview with Dr Greg Mills at the Brenthurst Foundation, Johannesburg, April 2011.

17 Interview with Niel Steyl, Former Executive Outcomes Pilot, Pretoria, April 2011.
18 Interview with Colonel Velthuizen (Retired) of the SADF, Pretoria, April 2011.
19 Interview with Niel Steyl, Former Executive Outcomes Pilot, Pretoria, April 2011.
20 Interview with Former Diplomat Sean Cleary, Cape Town, April 2011.
21 Interview with Professor Annette Seegers, Cape Town, May 2011.
22 Interview with journalist Tim Butcher, Cape Town, May 2011.
23 Interview with Defence Consultant Helmoed-Romer Heitman, Johannesburg, April 2011.
24 Ibid.
25 Interview with journalist Tim Butcher, Cape Town, May 2011.
26 Ibid.
27 Interview with Dr Greg Mills at the Brenthurst Foundation, May 2011.

6   American contractors

The contractor has become the poster boy for the Iraq War. Armed to the teeth and high on steroids, these private agents have earned themselves a reputation for being overly aggressive, negligent and unaccountable. This has inevitably led to tensions between the professional soldiers on the ground who are mandated with keeping the peace in Iraq and the private security companies who are 'in for a buck'. Nonetheless, the US government has increasingly turned to the private sector as an allegedly cheap and immediate solution to the shortage in military personnel that threatens their ambitions in Iraq: by 2007, the "Defence Department (had) spent $158.3 billion on services, a 76 percent increase over the past decade and more than what it spends on supplies, equipment and major weapons systems" (Isenberg 2009: 86). In 2009, 173,000 private contractors were believed to support the work of the 146,000 US troops on the ground. Out of these contractors hired by the Department of Defense (DOD), the State Department and USAID, 49 per cent were local Iraqis, 34 per cent were third-country nationals, and only 17 per cent were US citizens.

The emergence of private and foreign combatants as a key component of US military strategy has created new opportunities in US foreign policy, and exacerbated old tensions between the non-state actors and the professional armed forces. This chapter explores the rise of the private military sector within the context of privatisation and outsourcing that already characterised US domestic policies. Using Iraq as a case study for America's experiments in military public-private partnerships, it outlines the advantages and difficulties of hybridising the armed forces and focuses on civil–military relations, issues of command and control, overlapping identities and blurred hierarchies. The massive outsourcing of logistics, tech support and even military tasks in the last decade has changed the face of security in Iraq, with contractors quickly outnumbering regular troops. Military operations abroad are increasingly supported or even replaced by private military and security firms, leading to a difficult cohabitation between the two actors that can adversely affect the war effort. Contractors are here to stay, however, and therefore it is critical to thoroughly grasp the obstacles hindering good civil–military relations in order to overcome them and adapt to the new military landscape of hybridised armies.

The changing military landscape

The rise of the private security industry at the end of the twentieth century and the beginning of the twenty-first was unprecedented: by 2010, the industry had reached $200 billion in annual revenue. The emergence of private security and military companies on the global scene defies contemporary anti-mercenary norms. In their corporate forms, mercenaries operate freely on the open market, offering military and security services to individual governments, private investors and international non-profit organisations. The explanation for this sudden expansion of the private sector stems from (1) the new security environment in which state actors now operate; (2) the ideological changes and general risk aversion that characterises democratic societies; and (3) the culture of outsourcing and privatising state services that has been embraced by Western governments.

Military downsizing

The end of the Cold War and the fall of the Soviet Union shifted the military goals of most nations. With the removal of an imminent threat of war, the huge conventional armies of the twentieth century became redundant. This led to a large-scale downsizing in the standing armies of the major powers: combined military manpower dropped by 50 per cent from 6,873,000 in 1990 to 3,283,000 in 1997. In the United States alone, the active force in 2000 represented only 64 per cent of its 1989 total: "the US Army, Navy and Air Force all registered reductions of about 35% between 1989 and 2000, and the Marine Corps registered a 12% reduction" (Ortiz 2010: 52).

The downsizing of military personnel was not necessarily followed by a reduction in military spending. Between 1989 and 2003, US military spending was reduced by only 1 per cent, but has since gone up again with the invasion of Iraq and the ongoing wars in the region: "the George H. W. Bush administration's Fiscal Year (FY) 2007 budget request of $439 billion marks an increase of approximately 27% in real terms since September 11, 2001" (Isenberg 2007). On the other hand, most European countries have cut their defence expenditures since the end of the Cold War. France cut its defence spending by 10 per cent, Germany by 29 per cent and the UK by 21 per cent. This is partly due to the consolidation of the European Union and the creation of the Common Security and Defence Policy which increased cooperation between member states and has thus reduced the likelihood of a war developing within Europe. Austerity measures and the economic crisis in Europe has required further cuts, with the British "Ministry of Defence in the process of cutting 25,000 armed forces personnel and 29,000 civilian staff by 2015, in the biggest round of cuts to the military since the end of the Cold War" (BBC News 9/2/2012).

The demobilisation of armies at the end of conflicts has always been followed by an overwhelming supply of former soldiers with highly specialised skills who flock onto the job market: According to Carlos Ortiz, "in the case of the United States and the UK, there appears to be some correlation between the downsizing and the size of their private military industry". This trend was also visible in South Africa at the end of the apartheid regime in 1994: as the country tried to erase its racist history by dismantling the military organisations that had enforced the apartheid policies, it became a prime provider of private military companies and military services.

The end of the Cold War created a power vacuum, with the new Russian Federation politically, financially and militarily uninterested in supporting its former allies in Third World countries. Concomitantly, the United States and its Western allies no longer had an incentive to promote certain political actors, particularly in Africa. Deborah Avant argues that demand for military skills on the private market increased as Western states downsized their armies, developing countries upgraded theirs and weak and failing governments sought new avenues to support their regimes.

Ideological changes

The changes affecting modern armies today are a reflection of "large-scale social changes in the broader society" according to Charles Moskos. The end of the Cold War affected the perceived level of threat of an imminent outbreak of war. The survivability of the state was no longer in question, and armies and societies needed to find a new purpose for their soldiers. The wars of the 1990s and the beginning of the twenty-first century are typically civil wars taking place in the Balkans or in former colonies. Immediate threats of terrorism have been isolated incidents and do not require mass mobilisation. 'Geo-economic' threats such as uncontrolled immigration, transnational crime, environmental disasters and epidemics have superseded 'geopolitical' priorities in the national security agenda (Luttwak 1993). The armed forces have therefore had to redirect their operations and rebuild a new identity in "operations other than war", such as peacekeeping and humanitarian missions.

Naturally, the structure of the armed forces has had to adapt to this new level of threat perception. During the Cold War, European states generally relied on a system of conscription to staff large standing armies. The fall of the Iron Curtain, however, accelerated the shift towards a volunteer army: conscription in France was abolished in 2001, Germany suspended conscription in 2011, and only six European countries still practice a form of conscription in 2012. The United States had ended the draft in 1973 and has since depended on an ever-dwindling army of professional soldiers. According to Moskos, the progressive abandonment of citizen-armies is a phenomenon of postmodernity as "during the late 1990s, the decay of the public service tradition was evident, conscientious objection had increased and (France) was moving rapidly toward a smaller, more professional volunteer military focused on peacekeeping and humanitarian operations".

Public attitude towards the armed forces has also played a significant role in the changing military landscape of the twenty-first century. While during times of conflict, the public and media are generally supportive of their militaries, the lack of imminent threat and the geographical remoteness of active engagements have affected society:

> in the post-Cold War era, the public mood toward the armed forces becomes more one of indifference. The end of conscription makes military service less salient to the general populace. The likelihood of volunteer recruitment drawing upon future elites and opinion leaders becomes increasingly remote.
>
> (Moskos *et al.* 2000: 20)

Conscientious objection has become an accepted feature of modern states and is no longer punishable. This reflects a public apathy towards military service and to civic duty in general. The improvements in the standard of living and increased access to education have further discouraged voluntary enlistment as young men and women enjoy the luxuries of day-to-day life.

Moskos further argues that today's Western societies are also less able to accept casualties in war. Edward Luttwak explains that this is a characteristic of 'post-heroic' societies where war has lost its allure and no longer arouses enthusiasm among either the population or the soldiers. This change in attitudes stems from the "small family size of post-industrial societies (that) makes such societies' populations loath to suffer wartime casualties" (ibid.: 29). Moskos adds that "the willingness of a country to accept casualties is positively related to the proportion of elite youth who are putting their lives on the line" which in turn requires that the government in power articulates clearly the immediate threat to national security to encourage public support for the war effort. The aversion to casualties is exemplified by the ongoing body-bag count that defines the success or failure of military operations today. Public support for war naturally has an effect on domestic politics, particularly in democratic countries. Consequently, and in view of contemporary attitudes to risk aversion, governments have sought to protect their soldiers from exposure to lethal situations, creating a paradox that jeopardises military effectiveness for the sake of domestic support.

Although military values are still a pillar of most professional armies, the concept of civic duty has gradually been eroded in postmodern societies. Furthermore, the abolition of conscription and recent socio-economic changes have affected public attitudes towards the armed forces. While public opinion in most countries has remained supportive, citizens are generally opposed to military service in their own country. Governments, particularly in the United States, have had to take into account this change in mood when making foreign policies that require military participation.

Privatisation

Military downsizing and the aforementioned changing public attitudes have forced the US government to restructure their defence sector. A study by the Defense Science Board Task Force published in 1996 reported that "all DoD support services should be contracted out to private vendors except those functions which are inherently governmental or directly impact war-fighting capability, or for which no adequate private sector capability exists or can be expected to be established". The study concluded that up to US$12 billion could be saved annually if the Pentagon "contracted out all support functions except actual warfighting" (Isenberg 2009: 2). By 2005, "the federal government was spending about $100 billion more annually for outside contracts than on employee salaries" (ibid.: 15) and up to 80 per cent of the federal budget in select departments went straight to private contractors. The privatisation of services was further accelerated under Defence Secretary Donald Rumsfeld, who promoted the use of civilians for all non-combat jobs with the purpose of freeing up and concentrating military manpower and resources on the war on terror.

The practice of outsourcing government services to the private sector is a fundamental part of US history: for the American Revolution, 700 private ships and 30,000 Hessian mercenaries were commissioned to fight against the British. Civilians were also used for logistic support the Second World War, the Vietnam War and the Bosnian War. The contractor-to-military ratio has shrunk exponentially with civilians overtaking military personnel by 2010, as clearly shown in Table 6.1 on the estimated ratio of contractors to US military personnel in recent wars.

The end of the Cold War has shifted the focus from the population to the individual. Citizens in the United States and Europe have such excessively high expectations from their governments that outsourcing is inevitable in order to keep up with the populations' demands. Even foreign policy, once the exclusive domain of the state, has been privatised as "everything from diplomacy to development to intelligence" (Bobbitt 2009: 87) is being outsourced. According to Philip Bobbitt, the modern *market* state claims its basis for legitimacy on its ability to "maximize individual opportunity and by adopting methods of warfare and defense unavailable to the nation states". Privatisation has therefore become an intrinsic feature of the market state, with defence, security and even ambassadorial services being provided by private companies, albeit under the umbrella of a government contract.

Table 6.1 Estimated ratio of contractors to US military personnel in recent conflicts

War	Contractor numbers	Military numbers	Ratio
Gulf War	9,200	541,000	1 : 59
Bosnian War	1,400	20,000	1 : 14
Iraq War 2006	21,000	140,000	1 : 7
Iraq reconstruction 2010	173,000	146,000	1.2 : 1

Corporate armies

The rise of private military and security companies is a natural development in the market economy. The downsizing of armies forced an abundant supply of professional soldiers looking for employ, and equally created a new demand for private sector support in order for states to carry out their foreign policy. Recognising a potentially lucrative opportunity, former soldiers began to come together to form corporations offering their military and security skills on the international market. Their ability to successfully commercialise military skills is evidence of a significant change of norms regarding the state's previously exclusive claim over the legitimate use of violence.

Unlike the companies of mercenaries established in the Middle Ages, private military companies are legally established commercial enterprises. Some of them are even listed on the stock exchange, with ArmorGroup becoming the first independent, international security firm to be traded in 2004. According to Ortiz, these firms "are formally incorporated, and although not exactly paragons of transparency, they produce corporate literature, attend international conferences, maintain Web sites, and tend to be affiliated to defense or security professional associations". They also pay taxes, promote their services through marketing campaigns, and enter into contractual agreements with recognised governments, non-governmental organisations, and multinational corporations. PMCs claim to be experts in the "application and transmission of the knowledge of the use of force" (Ortiz 2010: 49). Services range from support and reconstruction to security, intelligence and even combat.

The largest private military companies are established in or around political capitals, particularly in London and in Washington DC (USA), and they employ a multinational team of professional soldiers and policemen, with recruitment efforts recently taking place in 'third-nation countries': contractors working in Iraq are former soldiers from the Balkans, Latin America, Easter Europe, New Zealand and Australia, Britain, Nepal, Israel and South Africa; many have also served in the French Foreign Legion. British company ArmorGroup has also employed hundreds of Gurkhas as guards for US firms Bechtel and Kellogg, Brown & Root. Former soldiers from Chile, Colombia, Guatemala and Nicaragua have equally been hired by American firms Triple Canopy, Inveco International Corp. and Blackwater. Officially, approximately 30 per cent of contractors in Iraq are third-country nationals. The services of PMCs are often transborder, with operations taking place simultaneously in different countries, depending on demand. These corporations are also exclusively profit-driven, leading to allegations of 'mercenarism'. Such accusations, however, are redundant in the twenty-first century where the principal employers of these firms are the US and the UK. Indeed, the biggest difference perhaps between "ragtag bands of adventurers, paramilitary forces, or individuals recruited in specific covert operations" (Isenberg 2009: 4) and organised clandestinely, and the modern corporate firm, is that the latter operates openly and freely with the overall blessing of the international community.

Although most private military companies claim to work only for legitimate governments, in Iraq, a majority of PMCs do not work directly for the US government as they are subcontracted from prime contractors to provide personnel protection to Iraqi and foreign companies seeking business opportunities in the country. The initial chaos that followed the invasion of Iraq in 2003, however, led to an environment of impunity and disinformation that undermined the reliability and reputation of corporate armies.

A case study of hybridisation in Iraq

Iraq has been an experiment in the use of private contractors in warfare. Between the beginning of the war in March 2003 and the latest census in 2011, the ratio of civilian contractors to army personnel operating in the field has ballooned from 1:10 in 2003 to 1.2:1. This disproportion between civilians and military personnel was an entirely new phenomenon which the new Coalition Provisional Authority, under the administrative leadership of L. Paul Bremer who was tasked with the reconstruction of Iraq, was unable and unwilling to control. The incidents that followed both marred the reputations of many private companies, undermined the war effort, and frustrated the US soldiers stationed in Iraq. Nonetheless, the reconstruction of Iraq could not have taken place without the added support and security that contractors offered to investors, governments and NGOs. The role of private contractors in Iraq has received enormous media and academic attention and offers unprecedented details on the interactions between civilians and military personnel in a war zone that can inform the debate on the effective hybridisation of the armed forces.

Invasion, insurrection and civil war

In 2003, the United States and its allies pre-emptively invaded Iraq with a combined force of 300,000 troops, including 46,000 British soldiers and a token contribution from Australia and Poland. The number of troops, however, amounted to less than half deployed in the first Gulf War. This was despite General Shinseki's estimate that a minimum of 500,000 soldiers were necessary to occupy and pacify Iraq, a country with a population of 30,399,572 spread over a surface area of 438,317km². Despite a quick victory over Iraq's conventional forces, culminating with President George W. Bush's premature declaration of "mission accomplished" on 1 May 2003, the allied forces soon found themselves short-handed, overwhelmed and unable to manage the levels of violence that continued to destabilise Iraq after the official cessation of hostilities. Security following the 2003 invasion was understandably at an all-time low, with reports of massive looting taking place, particularly of archaeological sites throughout Iraq, including the National Museum of Iraq. At least 250,000 tons of weaponry and ordnance which were also stolen would later fuel the insurgency. The armed forces focused their limited manpower on protecting hospitals, water plants, select ministries and oil refineries.

Despite a political end to the war, violence increased towards the end of 2003, with a surge in guerrilla attacks. The insurgency gained momentum in the spring of 2004 as foreign opponents to the occupation flocked into Iraq from neighbouring countries. An organised insurgency, allegedly led by Abu Musab al-Zarqawi, encouraged the targeting of Iraqi civilians and security forces with a series of massive bombings. This campaign of violence was meant to challenge the authority of the new transitional government and to demoralise Iraqi security forces and citizens. Concomitantly, a nationalist and Islamist Sunni insurgency and the Shia Mahdi Army also launched attacks on the coalition in an effort to force them out of the country. The widespread reach and planning of the attacks suggested an insurgent coordination against occupying forces. The coalition responded with a military counteroffensive, hunting down former leaders, and began the political handover of the country from the transitional Coalition Provisional Authority to an Iraqi government, with elections taking place in January 2005. A total of 140 US soldiers were killed during the invasion of Iraq, whereas 93 per cent of all US military casualties took place in the insurgency that followed. The US Army lost a further 849 soldiers in 2004, 823 soldiers in 2006, and reached an all-time high in 2007 with the death of 904 soldiers.

While the counterinsurgency effort proved successful, in 2007, sectarian violence between the Shi'a majority and the Sunnis began to tear the country apart, undermining the peace and reconstruction efforts of the Allies. Shi'a militia groups began to exert vigilante justice through zealous death squads while Sunni insurgents affiliated with Al Qaeda targeted Shi'a religious sites. The Al-Aksari Mosque bomb attacks by Sunni rebels in 2006 and 2007 sparked a spiral of violence as both parties increasingly targeted civilians and religious sites. The civil war further discredited the Allies' ability to restore the country, with the ongoing hostilities pushing 1.7 million refugees out of the country and forcing another 1.3 million persons into internal displacement by 2011. The continued violence also slowed down reconstruction efforts as aid workers, private individuals and contractors were targeted by insurgents who used terror tactics such as kidnappings and torture to exert political pressure on the government and discourage foreign interference.

Table 6.2 Civilian deaths by year

Year	Civilian deaths	
2003	12,087	
2004	11,152	
2005	15,491	Insurgency
2006	28,225	
2007	25,063	Iraqi Civil War
2008	9,385	
2009	4,713	
2010	4,045	
2011	4,087	

Although the United States and its allies invaded Iraq in 2003, they did not anticipate the subsequent insurgency and civil war: they believed that "reconstruction would take place in an environment with little threats from insurgents or terrorists" and had made "few or no plans for any other condition" (Petersohn *et al.* 2010). The extent of the unrestrained violence against the Iraqi civilians and security forces, the occupying armies and any foreigner who happened to chance into the country severely hindered the reconstruction project that was a priority in the US's plan to hand over power to the Iraqis and bring democracy to the Middle East. It emerged in hindsight that the Bush administration had "grossly underestimated the number of troops that would be required for stability and security operations" (Isenberg 2008: 157).

Casualty aversion

The violence began to slow down in 2009, with the election of President Barack Obama, who announced an 18-month withdrawal timetable for all US combat troops. A force of up to 50,000 American troops was to remain in Iraq to train, equip and advise the Iraqi Army and to support and replace withdrawing forces with the support of private military and security personnel contracted by the US government.

Although US casualties dropped after 2009, with the loss of 'only' 60 soldiers in 2010 and 54 soldiers in 2011, public support for an ongoing American presence in Iraq dwindled. The American public has always been critical regarding combat casualties according to a 2005 RAND report on American Public Support for US Military Operations. This was strongly demonstrated in the media outcry over the death and desecration of American soldiers in Mogadishu in 1993 that prompted the withdrawal of all US troops from Somalia. The US government, recognising the impact of body bags on public support, had banned media coverage of returning fallen soldiers in January 1991, at the instigation of then Secretary of Defence Dick Cheney. This was an important move, reflecting the firm belief that "inherent in our democratic notions of governance, public support (or public consent) is critical to any successful military action abroad" (Schooner, Swan and Collin 2011). The policy of hiding military casualties from public view was maintained by the George W. Bush administration during the Iraq War but was reversed by the Obama administration in February 2009.

According to RAND, 'casualty sensitivity' is proportional to the prevalent belief about the importance of the stakes and the probability of success (Larson and Savych 2005). As combat operations wound down in Iraq, the American public became increasingly impatient with the loss of their sons and daughters in a foreign land, where the immediate threat of a hostile regime had now disappeared. Peace and stability have never been a pillar of military operations, and did not justify the continuing exposure of US troops to high levels of violence. The insurgency and civil war were considered by many as a domestic matter, or at least an international responsibility, as were the reconstruction efforts.

Casualty aversion therefore was a driving force for the massive use of private contractors in Iraq, as their deaths did not feature in official figures. Consequently, the United States turned to the private sector to appease the American population, and to fill the logistical vacuum, awarding contracts of a total value of US$158,3 billion in 2007 to private military and security companies. By 2011, it was estimated that at least 173,000 private contractors were operating in Iraq. From these numbers, at least 1,537 men have been killed while carrying out their contract, of which 20 per cent are US citizens and over 65 per cent are Iraqi private contractors. Another 51,000 contractors have been injured. The US Department of Defense claimed that an average of "one civilian contractor is killed for every four members of the US Armed Forces" – but many of these men are not US citizens and none of them are included in the official body-bag count. A study conducted by Isenberg in 2011 found that contractor fatalities represented 45 per cent of all fatalities. The substitution of soldiers for contractors "has taken place outside of the cognizance of the public" and the 'sacrifice' of these men remains happily ignored by society (Schooner and Swan 2011).

Dependence on private companies

In 2003, Paul Lombardi, CEO of DynCorp, said "You could fight without us, but it would be difficult. Because we're so involved, it's difficult to extricate us from the process" (cited in Isenberg 2008: xii). Private military companies were present, if relatively unnoticed, during early combat operations. They were primarily hired by the US government to provide the US Army with essential logistical support:

> the military relied on civilian contractors to run the computer system that generated the tactical picture for the Combined Air Operations Center for the war in Iraq; (…) contract technicians supported Predator unmanned aerial vehicles and the datalinks they used to transmit information. The U.S Navy relied on civilian contractors to help operate the guided missile systems on is ships (and) the Army depends entirely on civilian contractors to maintain its Guardrail surveillance aircraft.
>
> (Ibid.: 24)

The US' policy of outsourcing non-combat or non-military jobs to the private sector took on new proportions

> as the Army was so short-handed it had to call up tens of thousands of reservists to fight in Iraq. Rumsfeld said he intended to assign the troops to military jobs and hire civilian workers or contractors to take the nonmilitary jobs.
>
> (Ibid.: 18)

Contractors now deliver at least 50 per cent of the US Army's logistical capacity. By 2009, private contractors had become an intrinsic part of America's war and reconstruction efforts in Iraq.

While private contractors entered Iraq initially with military support firms and military consulting firms offering training and advisory services, they gradually moved towards the battlefield. Worldwide Personal Protective Services (WPPS) contracts were awarded to the private companies DynCorp and Blackwater in 2004 to meet the security needs of the US Embassy in Baghdad. The Regional Embassy Office in Basra was protected by another PMC, Triple Canopy. Although these contractors were expected to perform only protective duties for high-level officials, they had to be well armed to carry out their mission.

Whereas the US government offers WPPS contracts to private companies to guard American public servants, foreign companies that are seeking to do business in Iraq and NGOs wishing to participate in the reconstruction efforts are forced to pay for their own security. Considering the high risk associated with conducting affairs in a war zone, protection is crucial and has driven the market for PMCs. Most private military companies, in fact, are subcontracted from primary providers to provide protection for businessmen. Only eight out of 60 or 13 per cent of the PMCs operating in Iraq were found to be hired by and therefore directly accountable to the Coalition Provisional Authority.

Reconstruction work is the principle activity of contractors in Iraq, however. Halliburton and its subsidiary Kellogg, Brown & Root have been the recipient of logistic contracts worth as much as US$300 million. Between 2003 and 2007, US agencies awarded US$85 billion to private contractors in Iraq, of which US$6–US$10 billion, or less than 10 per cent, was put towards security. Nonetheless, it is the security aspect of contracting that has received the most attention. The industry has been tainted by a lack of transparency and accountability, and by a handful of high-profile lethal incidents.

Legal immunity

For the first three years of the war, the United States had very little control over or information on the contractors working under the umbrella of the US government. Furthermore, most PMCs were employed by foreign companies, putting them beyond the reach of American authority. Indeed, since their arrival in Iraq, private security companies have earned themselves an infamous reputation, known for their 'licence to kill' and their immunity from the law. Scandals involving fraud, murder, torture and confrontations between the regular armed forces and the contractors have generally gone unpunished, testifying to a lack of control and legal accountability exerted over these private companies.

For the first year of the Iraq War, there was neither a legal framework nor any oversight to govern the actions of private contractors. It was only in June 2004 that the Coalition Provisional Authority, under the leadership of Paul Bremer, issued Order 17 on the "Status of the Coalition, Foreign Liaison Missions, their Personnel and Contractors". The Order bestowed immunity from "local criminal, civil and administrative jurisdiction and from any form of arrest and detention other than by persons acting on behalf of their parent states" to all US troops and civilian personnel, effectively putting them beyond the legal jurisdiction of the Iraqi authorities. In section 3, clauses (1) and (2) explicitly exempt all contractors and subcontractors, other than those "normally resident in Iraq", from "Iraqi laws or regulations" for all activities carried out both within the terms of their contract and "not performed by them in the course of their official activities" during the entire duration of authority of the CPA (section 4).

Despite this perceived blanket grant of immunity, contractors were only exempt from Iraqi laws and tribunals; they were not beyond the authority of the CPA, and were required to comply with the laws and regulations of the provisional government. Memorandum 17, signed by Bremer in June 2004, required that all private security companies "be registered, regulated and vetted" by the Ministry of Interior and Trade. On the other hand, the Rules for the Use of Force by Contractors in Iraq, annexed to Memorandum 17, clearly stated that "nothing in these rules limits your inherent right to take action necessary to defend yourself". Consequently, the legal provisions for contractors served both to emasculate Iraqi authorities and create an impression of control on the part of the CPA, while endowing private companies with a large margin of autonomy justified by a claim of self-defence. Furthermore, "both contractors and US government officials say that Memorandum 17 is impossible to implement (...) as the Iraqi government has neither the personnel (nor) the capacity to enforce the order" (Isenberg 2008: 144).

Since the official US handover of power in 2011, however, contractors have reported a reversal of fortunes, with alleged abuses by Iraqi bureaucratic personnel and security forces. Contractors and foreign private security companies have lost their immunity and are now faced with absurd amounts of red tape and ongoing harassment. In a letter addressed to Hillary Clinton in January 2012, the International Stability Operations Association (ISOA) lamented the difficult operating conditions: "the lack of visas or renewals is preventing our member companies from deploying into Iraq in support of embassy contracts and has led to the detention and expulsion of a number of member companies' employees". In addition, "approved movements have been subject to stops, detentions and confiscation of equipment without justification, impacting delivery of equipment, supplies, and materials to the US embassy, bases and offices throughout the country". Consequently, several private security companies have shut down operations in Iraq.

Fraud

There has been a considerable lack of accountability and oversight on the part of the US government with regards to the private security sector: during the first three years of the war, there was a dearth of information and the US had no accurate count of its contractors even though more than US$766 million was spent on private security through the end of 2004. Because of the urgent circumstances in Iraq, many contracts were allegedly awarded to companies without competitive bidding, demonstrating the preference for personal relations over democratic processes. This process inevitably eliminated any opportunity for free-market control, which would have arguably held firms accountable for their actions due to the competitive nature of the free market. The lack of transparency in the allegation of security contracts had also led to allegations of "cost overruns, inflated invoices, fraud, and abuse" (Isenberg 2008: 161).

The opportunity for fraud was illustrated in the case of Custer Battles, a private security company that was accused of cheating the US government out of up to US$50 million. Custer Battles landed a US$16 million contract with the CPA to guard Baghdad International Airport and a second contract to provide logistical support for a currency exchange programme. The company was subsequently accused of creating false invoices, inflating the 25 per cent mark-up to 162 per cent and billing the US government for it. Allegations of misconduct included knowingly providing sub-optimal equipment. In the testimony, the airport's director described Custer Battles as "unresponsive, uncooperative, incompetent, deceitful, manipulative and war profiteers" (Isenberg 2008: 88). Custer Battles was found guilty in a federal court of "defrauding the United States by filing grossly inflated invoices for work in the chaotic year after the Iraqi invasion" (Eckholm 2006). The ruling was overturned on a technicality, however, not because the company was absolved of having committed a crime, but because the CPA could not be considered a US agency. Custer Battles was also one of the companies whose employees allegedly fired upon Iraqi civilians.

The scandal of Custer Battles began to symbolise the poor monitoring and accountability system of the private security industry in the months following the Iraq invasion. The private security company was essentially given US$16 million in a no-bid contract, having demonstrated little if any experience prior to their Request for Proposal. The CPA "forked over $2 million anyway, in cash, to get them started" (Isenberg 2008: 91). The Commission on Wartime Contracting found that "war planners have wasted as much as $60 billion on contract fraud and abuse in Iraq and Afghanistan, about $1 for every $3.50 spent on contractors in those countries over the last decade" (Shane 2011). Out of this amount, between

> $21 billion and $42 billion of the lost money has gone to incomplete construction and training projects, unnecessary subcontractors, unexplained cost overruns and similar wasteful practices (and) another $10 to $18 billion has disappeared due to fraud, including bribes to local government officials and contractors who simply ran off with thousands of dollars

according to the 2011 Commission on Wartime Contracting. The Defense Criminal Investigative Service had received more than 1,500 complains for fraudulent activities in Iraq and Afghanistan. The report concluded that the government simply did not have the capacity to oversee the number of contractors that were operating in Iraq, and had awarded too many contracts without a proper bidding process, undermining the notions of competition and free market economy.

Civilian casualties and abuses

Private contractors are accused of having been complicit or perpetrators of ethical and legal abuses against civilians. These incidents have undermined the counterinsurgency operations of the US Army by fuelling local resentment against the foreign occupiers who refuse to hold accountable those responsible for the violence. Evidence of gross misconduct on the part of private security and military companies have caused international outrage and affected the relationship between the armed forces and the civilian contractors. There are no official statistics, however, on the number of incidents instigated by private contractors against civilians in Iraq, although a number or organisations have attempted to compile this data. More than 4,500 pages of incident reports were released by the Bureau of Diplomatic Security in 2008, and describe about 600 incidents where security contractors have fired a weapon. Four hundred thousand US military documents were also published by the whistle-blowing website Wikileaks, with details of operations in Iraq between 2003 and 2009, and have served as an imperfect source for research into this shadowy business. Approximately 65 per cent of reported incidents described contractors firing their weapon into a civilian vehicle, but with little if any details regarding the consequences of these actions. Journalist Robert Young Pelton claimed that "in my time with contractors in Iraq, I never saw a single report filed, even though gunfire against civilians was an everyday event, possibly to an average of three to six warning shorts per run". Most incidents were never investigated by the authorities. Ten Iraqi deaths and 14 incidents of injuries were recorded in these files. Half of all incidents reported took place outside of Baghdad, although misconduct is less likely to be recorded in the provinces as there are fewer witnesses.

A highly visible and publicised incident of contractor misconduct took place on 16 September 2007, when five contractors from the already infamous private military company Blackwater fired upon civilians, killing 17 and wounding 24 Iraqis in Nisour Square. Although the contractors claimed self-defence, this was denied by Iraqi officials. Nonetheless, Blackwater continued to operate in the country, and nobody was held accountable for the unlawful killing of the Iraqi civilians. This contributed to fuelling local outrage and resentment and has "undermined the military's war for the hearts and minds of local civilians by failing to abide by the laws and norms of just war" (Dunigan 2011: 45). Despite the high profile of this incident, it was not by far the first occasion where Blackwater had been accused of misconduct: on Christmas Eve 2006, a drunk contractor from the firm "shot dead the security guard of the Iraqi Vice President" (Singer 2007); in February 2007, Blackwater contractors shot and killed three security guards at the Iraqi Media Network in Baghdad; and in May 2007, "guards opened fire on the streets of Baghdad twice in a single week. In one incident a guard shot and killed an Iraqi driver near the

Interior Ministry" (Isenberg 2008: 141). Despite these incidents, the State Departments extended its contract with Blackwater in September 2009. This was eventually revoked and Blackwater, then Academi, was banned from operating in the country.

Although Blackwater has received much negative publicity for its excessive use of violence, several other private security and military companies have been accused of similar misconduct: employees of London-based defence contractor Aegis "posted videos on the Internet in 2005 showing company guards firing automatic weapons at civilians from the back of a moving security vehicle" (AP 12/8/2007). In another high-profile case, two prominent companies, CACI and Titan, were both implicated in the notorious Abu Ghraib prison scandal where detained Iraqis were tortured, abused and killed. The US Army claimed that contractors were implicated in 36 per cent of all proven abuses in Abu Ghraib, and accused six employees of being 'individually culpable'. No civilian contractor, however, faced prosecution, whereas several US military were tried and punished. Furthermore, the involvement of Titan and CACI in the scandal did not negatively affect their business: Titan was subsequently awarded multiple contracts worth several million dollars.

Allegations of contract misconduct in Iraq have largely been buried in the chaos of war, where suicide attacks and car bombings are a daily occurrence. Contractors operate in a high-stress environment, with little legal accountability but huge professional responsibility vis-à-vis their employer and their clients. Contractors also have a higher proportion of casualties (10 per cent) per reported incident, compared to coalition forces (0.79 per cent). Although this does not justify misconduct, it does give a context for the environment in which these incidents happen. The armed forces have also been accused of operating with negligence: the Abu Ghraib abuses, for example, were largely led by US personnel. A significant difference, however, is that military personnel have been selectively held accountable for their actions, whereas contractors have largely gone unpunished. The impact of Abu Ghraib was notably for US interests in Iraq: General Mattis, commander of the United States Central Command, explains that "when you lose the moral high ground, you lose it all" (Mattis cited in Ricks 2006: 291). Contractor or soldier, the actions of US employees in the Iraqi prisons resonated around the world.

Blue-on-white incidents

'Blue-on-white' is the term used to describe 'friendly fire', exchanged between military personnel and contractors in Iraq. This exchange is generally accidental, reflecting a lack of coordination, communication and trust between the two actors. Human Rights First argued that while 610 incidents were reported between July 2004 and April 2005, this probably reflected a fraction of the actual incidents that took place in that time frame. A RAND report published in 2010 found, however, that "the vast majority of reported blue-on-white incidents in Iraq are actually perpetrated by coalition forces against private security contractors, with most occurring when contractors are approaching checkpoints or passing military convoys" (Cotton *et al.* 2010).

The most famous incident of blue-on-white occurred in May 2005, when a 19-man convoy from US firm Zapata Engineering was arrested by the American Marines for firing upon them at a checkpoint watchtower. Sixteen contractors were remanded in US custody for several days as a result, and reportedly harassed and humiliated during their detention. Despite accusations of firing on US troops, the employees were never charged for any offence (Dunigan 2011: 63). Another private military company, Triple Canopy, also reported several blue-on-white incidents where US military personnel allegedly opened fire on them.

The prevalence of blue-on-white incidents in Iraq has not been reduced, despite reports of improved coordination between military personnel and security firms according to a Government Accountability Office (GAO) report published in 2008. Many incidents of friendly fire are blamed on the 'trigger-happy', uncontrollable contractors, even though contractors are the most likely victims of friendly fire, according to David Isenberg. Private contractors are not required to wear standardised uniforms, and many travel in unmarked vehicles at high speed, making them suspicious to army personnel in an environment of extreme volatility and violence. Not only is it difficult for the military to identify contractors as friendly forces, the lack of military–PMSC coordination, compounded with an inherent distrust between the two actors and a vehement disinformation campaign run by the international media, have led to a difficult coexistence in Iraq.

Analysis of civil–military relations

The rapid transformation from relying on military personnel to appointing security positions to civilian contractors has blurred the line between civilian and combatant and has led to a myriad of problems, including "issuing IDs and weapons permits; chain-of-command ambiguity, contrary objectives, coordination of security convoys and friendly fire incidents, not only from coalition troops firing on contractors mistaken as potential insurgents, but also between contractors and other contractors" (Pelton 2007: 107). The difficult cohabitation of American private contractors and US military personnel in Iraq has been the product of jealousy, misinformation, a vacuum of accountability and sub-optimal efforts between the two players towards improving conditions for coordination and communication. Furthermore, reckless behaviour on the part of the contractors and the companies that hire them have inevitably led to tensions between the professional soldiers mandated with maintaining the peace in Iraq, and the private, armed contractors who circulate freely inside the country. Overlapping identities, diverging salaries and a lack of military hierarchy and judicial accountability have consequently shaped the relationships between soldiers and contractors in the war in Iraq.

Identity and motivation

Of the estimated 161,000 private contractors working in Iraq in 2008, only 17 per cent, or 27,370, were US citizens. On the other hand, "all but 1,000 of the State Department's 5,500 contractors were US citizens" according to Dunigan. Although there is now a majority of Iraqi or third-country nationals working for private security companies in Iraq, the most sensitive missions outsourced by the Pentagon are generally awarded to Americans, who are considered to be the most trustworthy, loyal and competent contractors.

The invasion of Iraq in 2003 triggered a sort of gold rush for American ex-military personnel who could expect to earn US$600–US$1,000 per day handling security contracts in the country. Most of these highly paid contractors were former SEALS or Green Berets with significant experience, training and service behind them. Many of them claimed to be motivated by patriotism. Military sentiment, surveyed in the aforementioned 2010 RAND report, suggested that contractors must "have a sense of patriotic duty, since many are prior military". Evidently, patriotism is not the only motivation, with most contractors lured to Iraq by promises for quick riches: "money is obviously a strong incentive for contractors to work in a combat zone and separate from their families when they do not have to" (Dunigan 2011: 66). These 'patriots for profit' as labelled by Thomas Bruneau, are perceived as loyal combatants, who, due to their former military training, are not likely to abandon their posts when faced with imminent danger. On the other hand, they are neither motivated by honour nor recognition, since their function as contractors denies them the acknowledgement that is reserved for military personnel.

The reasons that drive an American to enlist in the armed forces or enter a contract with a private security and military company are not hugely divergent. The camaraderie, lifestyle, idealism and patriotism have lured these men to Iraq, either as soldiers or as contractors. Money has also been a source of motivation for young men and women to join the army, as has the promise of US citizenship, evidenced by the US Army's policy of endowing citizenship to immigrants who serve in the armed forces. Despite their many similarities, contractors were often subject to resentment from the part of American soldiers because of their perceived pay differential, their immunity from the law and their carefree undisciplined attitude that loudly illustrated the differences between these armed combatants.

Pay

There is a significant difference in salary between contractors and military personnel: in the early days of security contracting in Iraq, "contractors (could) make nearly a soldier's annual salary in just one month" (Dunigan 2011: 63). RAND's report on the *Views About Armed Contractors in Operation Iraqi Freedom* found that interviewed military personnel cited pay differential as a major source of resentment and "was detrimental to morale". One US Army major explained that "There was always resentment between my soldiers and the contractors because the contractors were making several times more money than my soldiers and doing basically the same job".

The difference in payment is also reflected in the differences in lifestyle that contractors enjoy in Iraq. A contractor working for Triple Canopy described his experience working for a private security company, and how it compared to his former job as a Special Forces officer:

> the guy in the military is making $40,000 a year working twelve-hour shifts seven days a week while the contractor is making $240,000 a year and working four- to six-hour shifts four to six days a week. The guy in the military sleeps in a room with ten other guys while the contractor has his own trailer with air-conditioning, television, a DVD player, and a fridge. The military guy is under the control of the military, so no sex or drinking – the contractor isn't. The military guy wears a uniform while the contractor wears whatever he wants. The military guy stays seven months and is forced to stay longer while the contractor can leave whenever he wants.
>
> (Dunigan 2011: 64)

These differences in lifestyle reinforce identity cleavages and encourage a hostile environment fuelled by inequality and jealousy.

Isenberg argues, however, that the perceived pay differential between contractors and military personnel is not accurate. He cites various military studies to show that the estimated annual average compensation for Navy personnel was US$94,900 according to a 2004 Center for Naval Analyses report, whereas the Government Accountability Office calculated that compensation for active-duty service members averaged US$115,500 (Isenberg 2008: 33). Furthermore, military personnel enjoy better job security, health-care benefits and retirement privileges. A 2008 Pentagon Review of Military Compensation concluded that "military personnel's annual compensation meets or exceeds the 80th percentile when compared to their civilian peers of like age and education". Ann Jocelyn from the war lifestyle publication *Serviam Magazine* claims that after taxes, a contractor can earn on average US$95,700, which is still "38 percent more than the sergeant", albeit without health care, housing or retirement contributions which, once taken into consideration, leaves him with US$38,306.

Nonetheless, the perceived salary difference between soldiers and contractors has allegedly affected the armed forces, with a significant portion of highly trained military professionals leaving service for the private sector. The Pentagon responded by offering bonuses in 2005 of "of up to $150,000 to keep elite commandos, such as Army Green Berets and Navy SEALs, in the military and prevent them from being lured away to higher-paying jobs by private security contractors" (Isenberg 2008: 59). The British Ministry of Defence, on the other hand, offered its soldiers the option of taking a sabbatical during which they could work as contractors in Iraq for up to a year and then return to the army at the same rank. A US Army officer summarised the problem that this talent-drain created for the armed forces:

you get stuck with the training and security-clearance costs; the soldier lured to the private sector gets his salary doubled or tripled – then the contractor adds in a markup for his multiple layers of overhead costs and a generous profit margin, and bills the taxpayers. How is that cheaper than having soldiers do the job? The scam-artists tell us that using contractors saves money in the long run, since their employees don't get military health care and retirement benefits. But the numbers just don't add up. Contractors are looting our military – while wrapping themselves in the flag.

(*New York Post* 6/10/2007)

It costs the state eighteen months and US$257,000 to train a Green Beret. The attrition of military personnel is an onerous enterprise that penalises the effectiveness of the armed forces and poses serious challenges for the future.

Cooperation and communications

Iraq is the first conflict where contractors and military personnel have been operating side by side in overlapping functions. There has consequently been no precedent to inform the actors on the best manner to communicate and coordinate with each other. Part of the problem, according to the US Government Accountability Office, is that "US forces in Iraq do not have a command and control relationship with private security providers or their employees", therefore communication takes place in an informal setting where personal relationships determine the level of information. The lack of knowledge regarding military positions and the movement of contractors in a unit's area of responsibility has been a source of confusion and has led to unnecessary danger on both sides: "this was mainly due to the lack of interoperable radio and communications systems between the military and PSCs" (Dunigan 2011: 68). RAND found that security contractors and military personnel exchanged cell phone numbers on an *ad hoc* basis when they knew that they could be working in the same area of responsibility. Although this cooperation is not institutionalised, it nonetheless contributes to the integration of civilian contractors and the US Army. Nonetheless, the lack of standard operating procedures in place in Iraq has sometimes left contractors at the mercy of insurgent attacks.

The US armed forces, however, are not briefed in their pre-deployment on the role and positions of private contractors in Iraq, furthering an environment of suspicion, obstruction and jealousy. Furthermore, the tactical aims of contractors are to protect the 'principal' that they are guarding, with little or no regard to the local population or to the rules of engagement by which soldiers must abide. The often disrespectful attitude of private security guards has soured the Iraqi population towards the contractors who are indistinguishable from American soldiers. This has undermined the US's COIN campaign for the hearts and minds of the population. Molly Dunigan argues that

while the US and coalition forces recognize the value of cultivating good relationships with locals in a counterinsurgency – focusing largely on the hearts-and-minds approach in these two theatres – PSC personnel in many reported cases have harmed their own and the military's relations with locals.

This has naturally had a negative impact on military effectiveness in Iraq.

To address the lack of cooperation, regional Reconstruction Operations Centres (ROCs) were established to improve communication and coordination between army personnel and private security forces. These ROCs were managed by the British private military company Aegis, under a contract outsourced from the Department of the Army (DA). The PMC acted as a clearing house, tracking contractors, providing daily intelligence services and filing claims and reports of misconduct. Aegis' efforts were severely limited though, as two of the largest PMCs in Iraq, Blackwater and DynCorp, refused to participate, arguing that they were monitored separately by the US government. A 2006 GAO report claimed, however, that "private security providers continue to enter the battle space without coordinating with the US military, putting both the military and security providers at a greater risk for injury". The continued incidents of friendly fire undermine the integration and effectiveness of both the armed forces and private contractors and exacerbate the feeling of resentment between the two. Nonetheless, the increased exposure of the US Army to private security companies and their developing interdependence is leading to a shift in attitudes, which was illustrated in RAND's 2010 report.

Efforts to improve cooperation are obstructed, however, by a chain of command that does *not* overlap between the armed forces and private contractors. A US Army colonel explained that security contractors

> did not fall under my command. Normally they were attached to another command moving through the area or worked for [the US State Department].... The main issue is that they did not have to answer to me, so I had very little leverage over them.
>
> (Dunigan 2011: 74)

In one anecdote, US diplomat Richard Holbrooke was travelling in Iraq under private protection, and asked the driver of his vehicle to slow down in view of the high level of civilian traffic on the road. The contractor refused on the grounds that the officer had no authority and that his responsibility was to protect his passenger.[1] Private security companies are answerable to their shareholders, and to the contracting party, but not to the legal or military authority of the Iraqi or American governments. Military commanders therefore do not realistically exercise effective control over private security contractors.

Accountability

The lack of oversight from the Department of Defense has not only created opportunities for fraud and abuse, but has also undermined the democratic control of violence in Iraq. Private military companies are not bound by the rules of engagement like the armed forces, and therefore have a generous margin to exercise any behaviour they judge appropriate for the accomplishment of their mission. There has been no doctrine, very little contract management and no command and control of contractors. There are also no institutionalised disciplinary measures that could be brought against private security companies, other than firing the company or terminating the contract.

Outsourcing military services to private companies is a convenient practice for the US government and the armed forces. It enables them to absolve themselves of direct responsibility when things go wrong. While companies are not held accountable, incidents can be blamed on private individuals, and contracts can be terminated or replaced without damaging the overall reputation of the coalition in Iraq. Furthermore, as contractor deaths are not included in the official body count, the US government has a wider margin to manoeuvre in Iraq without attracting the disapproval of the American electorate. Dunigan criticises this undemocratic approach and explains that "in cases of co-deployment, PSCs act as force multipliers whose deaths do not have to be reported in official casualty statistics, thus allowing policymakers to deceive the electorate in such a manner as to disrupt the electorate's conflict selectivity". This in turn undermines democratic accountability as policymakers may choose to engage in a conflict regardless of public opinion, since the citizenry are not expected to contribute to the war effort.

The US government has deliberately outsourced important contracts of a military nature to private companies as a response to the logistical limitations imposed on them by a democratic society. Contractors have added value *because they are not held accountable*. Should it choose to do so the American government could create and implement legislation to discipline private companies and their rogue personnel. At this point, however, it appears that contractors are a valuable asset in the reconstruction and securitisation of Iraq, and bringing outliers to justice requires a significant monetary, legal and political investment that is counter-productive to US interests domestically and in Iraq.

Conclusion: hybrid soldiers in Iraq

In Iraq, the US government has instigated a policy of hiring private contractors to take over certain jobs previously held by military personnel and recruit private security and military companies to support the professional army in combat zones. Contractors have been used as interrogators in prisons, as guards in front of military bases, as consultants training Iraqi police and as bodyguards for high-profile Americans including the head of the Coalition Provisional Authority, Paul Bremer. The use of civilian contractors for military – and not only supportive – purposes has led to a new permeability of roles that blurs the traditional strongly defended line separating soldiers from civilians. Although this policy has created unforeseen problems regarding command and control, accountability and coordination, it has also planted the seeds of the armies of the future, where civilians and soldiers are co-deployed with the intent of complementing each other's strengths and weaknesses, depending on the budget of the state and the nature of the mission.

America's policy of outsourcing traditionally military functions to private contractors reflects a significant shift in norms. Soldiers are no longer the only accepted armed defender of a country's interests. The new security environment, the budgetary constraints of 'democratic' armies and society's aversion to sacrifice and military casualties have propelled forth a new actor: the private security company, with its armies of weapon-bearing civilian contractors.

Non-state actors have been a continuous presence in warfare throughout history, but they were generally foreigners. For at least the past two decades, the international community has sought to delegitimise mercenaries from warfare. The particularity of the war in Iraq, however, is that the mercenaries were operating private contracts on behalf of the US and its army, and that so many were also US citizens. Most of these contractors were equally veterans of the US Army. They share a common military and cultural background with the soldiers fighting in Iraq. While contractors are handsomely compensated for the risks they take, these men have also claimed that patriotism is a principle source of motivation for their presence in the warzone. Consequently, there is more common ground between the American contractors and military personnel than ever before in the history of mercenaries, in terms of their motivation, background, military training and cooperation efforts.

Despite their similarities, however, there is still huge discord and resentment between the two actors. Pay differentials are a principal source of envy, as is the freedom and lack of accountability that contractors appear to enjoy. Although contractors and soldiers share a common background, the structural environments in which they operate are completely different. Soldiers are subject to a relentless regime of training, function within a strict hierarchy, and are accountable for their actions. They develop an *esprit de corps* and functional ideology that is grounded in their common hardships and sense of responsibility towards the army, towards each other, and towards their country. Contractors, on the other hand, have a corporate operational culture with very little hierarchy, no accountability and no compulsory training. Although there appears to be a certain pride associated with working with particular private military companies – such as Blackwater which claims a distinct and superior culture to the other PMCs – there is no evidence of a deeper purpose or commitment other than to accomplish one's mission and pocket the salary. Dunigan concludes in her study that

> PSC personnel and the military are defined by quite different constitutive norms and social purposes in a war zone. For instance, two of the GAO's military interviewees noted that differences in operational cultures and missions made it difficult for the two groups to coordinate. Not only do PSCs vary among themselves in terms of operational styles, they also differ from the military in this sense.

Furthermore, the irresponsible and provocative behaviour of some contractors have severely damaged their reputation. They have been labelled as "cowboys", "mercenaries" and "pseudomercenaries", reflecting the disdain of the armed forces and the ongoing normative obstacles to the use of civilian contractors in military operations.

US Colonel Bill Gallagher argues that the integration of private contractors into combat roles is a result of changes in contemporary warfare where the dividing line between combat and peace zones is disappearing. Likewise, the division of functions between contractors and soldiers is being progressively dissolved. Although private military companies claim a purely defensive role, their personnel have found themselves in the heat of battle in the process of carrying out their contract. Blackwater guards, for example, have allegedly had to fight back an insurgent attack on the CPA headquarters in Najaf, and there have been rumours of DynCorp personnel supporting the US Army in impromptu combat operations. Notwithstanding their participation and operational support in the Iraq War and the high casualty rate, contractors are still being harassed by their military counterparts.

The weakening division between military and civilian leaves the soldier searching for his space and identity in the twenty-first century. Contractors are invested with an authorisation to use lethal violence, which was previously only entrusted to 'servants of the state' who had submitted to military discipline and the recognition of the laws of war. The collective identity of the armed forces and their role in society is arguably being eroded by the increasing involvement of private civilian contractors in the military theatre. Although public support for their national armies is at an all-time high, willingness to serve in the armed forces is dwindling, governments are under pressure to avoid casualties and the US Army is suffering from the demystification of the values of patriotism and sacrifice. Democratic states are increasingly dependent on private security and military companies to carry out their military ambitions. The impact on the identity and effectiveness of the armed forces, however, is quietly being left for the future.

Note

1 Anecdote provided by Professor Christopher Coker.

7 Citizens, soldiers and state control

Mercenaries are different from soldiers because they are not state actors. This means that they exist outside of the political realm and are not subject to state conditioning: "they are a band of people whose appetites are untrammelled by a sense of public spirit" (Coker 2010: 145). They are also beyond state control and accountability, marking them as potentially dangerous actors, particularly when mandated with the security and defence of a state.

The mistrust and fear of mercenaries are rooted in the social norm that the state, as the guarantor of peace and security, holds an exclusive monopoly over the legitimate use of violence. It exerts this privilege through its chosen servants, the soldiers which the state recruits and trains for this express purpose. Any foreigners to this established system of power are viewed with suspicion and censure. The choice of combatant reflects a society and its state's perpetually evolving normative values: "sometimes citizens are called upon to fight; at other times the state raises professional armies; and sometimes it contracts out to others" (ibid.: 149). The use of contractors, mercenaries and private military and security companies by the state in the twenty-first century suggests that a historic change in norms is taking place: the state still claims the prerogative to administer and sanction violence, but it no longer relies solely on state actors as it delegates part of its military mandate to external actors. Indeed, the case studies of France, Angola and Iraq in the preceding chapters show that non-state actors and national armies are increasingly being called upon to work together to carry out a state's foreign policy.

This chapter explores the changing role of the state and its reliance upon armed combatants to establish legitimacy and hold on to power. It analyses the use and manipulation of norms by the state, and shows how norms can be adapted to suit the purposes of the ruler, particularly regarding the waging of wars. As the state's need for combatants exceeds its domestic supply and capabilities, it has turned to foreigners, mercenaries and corporations to meet its demand, and has easily justified this shift to its citizens. By outsourcing its military needs to non-state actors, however, the state has reached a new zenith of power, where it can shift moral and legal responsibility away from itself and evade the system of democratic control that the national army claims to safeguard. The chapter concludes that mercenaries that are hired by a state uphold the power and centrality of said state, despite accusations to the contrary. On the other hand, they threaten established norms of democratic accountability by giving the state the ability to act beyond the will of the people, and undermine the morale and ultimately the strength and performance of the national armed forces.

State control

Security is at the very heart of human communities. No economic, personal, or political activity can successfully take place in an environment of recurrent existential threats. Thomas Hobbes claimed that the state came about to protect people against the dangers of the "state of nature": individuals were forced by the violent nature of mankind to forfeit a part of their freedoms and contribute to the safeguard of their community in order to improve their chances of survival in a brutish and anarchic world. "Every state is founded on force", declared Leon Trotsky in 1918, implying that, although other actors have wielded force in the history of mankind, only the state has successfully established its legitimacy based on its ability to control the level of violence within a given territory and therefore protect the community from itself and others. Max Weber concludes, therefore, that the state "is considered the sole source of the 'right' to use violence". Weber's definition of the state clearly assumes that the potential to use violent means and legitimacy are inseparable. This norm, regardless of whether or not it represents the actual abilities of all states in the international community, is nonetheless the standard for all states: modern states are expected to be able to control the use of violence within their territory and subsequently hold accountable any actor who threatens the state or challenges its authority.

Early predators and the princely state

Historian Charles Tilly famously claimed that: "war made the state and the state made war". The state structure emerged out of "a centuries long process of internal pacification and international warmaking" (Owen in Mabee 2011: 23) during which various warlords sought to control people and their lands in order to levy taxes and increase their own wealth. In the early Middle Ages, raids against the population were used as the main source of 'state' income. This required the collaboration of armed men who could intimidate and overwhelm any resistance. Chapter 2 argues that, in the absence of standing armies, bands of mercenaries and knights were the de facto choice of combatants to help a leader plunder and coerce the local inhabitants. Knights were the medieval equivalent of modern soldiers: men indentured to a leader who commanded their loyalty, trained them in the arts of war and led them in battles and raids; mercenaries were armed combatants temporarily hired to participate in a raid or conflict. The main differences at the time between knights and mercenaries lay in their personal allegiances and permanency. The use of knights, peasants and serfs to supply the armies of sovereigns shows an early preference for 'domestic' actors – i.e. men tied to their vassals by bonds of honour and economic dependence.

Leaders eventually realised, however, that raids disincentivised peasants from producing more than the bare necessities to survive, which limited the amount that they could in turn appropriate. Consequently, they shifted to a system of racketeering, or "stationary banditry" where leaders could levy a "predictable tax that takes only a part of his victims' outputs, thereby leaving them with an incentive to generate income (…) which might well increase output and tax receipts by a large multiple" (Olson 2000: 8). This tax was extracted by means of tribute, rents, dues and fees. The leader therefore became a sedentary ruler but in turn had to protect his territory from external raids and attacks, making him increasingly dependent on his army of knights that had to be permanently mobilised, trained and fed.

Once in power, rulers relied on violence or the threat of violence to consolidate their reign and legitimise their exclusive right to levy taxes upon the population. Initially, taxes were a means of payment for the security that the new state promised its subjects. Mancur Olson compared the sedentary state to a criminal organisation that seeks to maximise its self-interest by increasing taxes and providing public goods. While this may appear beneficial in the long run, the state's justification for levying taxes in the first place is to sell protection "both against the crime it would commit itself (if not paid) as well as that which would be committed by others (if it did not keep out other criminals)". This exchange nonetheless solidified the "contract between a ruler and the ruled through the trading of protection in return for other services like taxes, revenue and labour; i.e. the economic base of the state" (Tilly in Small 2006). These taxes were subsequently used to finance an army with which to enforce the state's authority within and beyond its legitimate territory: "the financial means thus flowing into this central authority maintain its monopoly of military force, while this in turn maintains the monopoly of taxation" (Elias in Tilly 1992: 85). Brian Mabee explains that

> one of the factors in the consolidation of states was the unintended consequence of the rulers' search for power – that the need for capital to invest in war-making inadvertently led to the elimination of rivals within a given territory, in order to have a greater capacity to extract resources.

The sedentary state thereby became the leading political actor as it managed to accumulate more funds, by means of taxation in exchange for guaranteeing protection, than any individual or group could achieve through pillage. Machiavelli's political environment emerged from this system where families and princes sought to "enhance their authority and security by promising those living under their authority security from an attack from outside forces" (Bobbitt 2002). The princely state, as Bobbitt calls it, existed to serve the interest of the prince whose power increasingly depended on the loyalty and servitude of its subjects and on its relations with other city-states.

By the seventeenth century, states had accumulated enough power to further consolidate their rule over the population by eliminating any potential for internal dissent. This was achieved by "general seizures of weapons at the end of rebellions, prohibition of duels, controls over the production of weapons, intro-duction of licensing for private arms, (and) restrictions of public displays of armed forces" (Tilly 1992: 65). In 1689, the English Bill of Rights prohibited the "raising or keeping a standing Army within the Kingdome in times of peace unless it is with consent of Parliament". Private armies were dissolved and the potential powers of individual citizens were severely controlled legally and militarily by the state. This Bill was challenged, however, by the Jacobite Rising between 1689 and 1746 during which Jacobite supporters were able to defy the Protestant rulers of Britain with an army of 8,000 men, made up largely of Scot-tish Highlanders. In the wake of the rebellions, Highlanders loyal to the crown were recruited and stationed across the Highlands to deter clans from assisting the Jacobites. These Highlanders were later integrated into the Black Watch Regiment which is still active in the British Army today.

By disarming the population, the state successfully claimed a monopoly over the use of violence within its territory. The right to use force in the name of the state was delegated to its chosen agents, who subsequently became state agents with a mandate to uphold the peace, and thus established the legitimacy of the state as the sole legal and able guarantor of security.

Monopoly of violence

The state has risen to power, maintained its legitimacy and projected its ambi-tions mainly through military means. Otto von Bismarck expressed his vision of state-building in his 1862 speech to the Landtag's budget committee: "not through speeches and majority decisions will the great questions of the day be decided – that was the great mistake of 1848 and 1849 – but by iron and blood". The modern state as it exists today was built from *Eisen* und *Blut*, with state for-mation and legitimacy historically attributed to military victories and defeats facilitated by access to cutting-edge technology. The state's monopoly of viol-ence is a prerequisite for the state to exert its rule over its population. The state also derives a significant proportion of its legitimacy from its ability to safeguard the lives and well-being of its citizens for whom it bares full responsibility. The reliance of the state structure on its armed force had naturally led to a symbiotic relationship where, in theory, the success of one is dependent upon the strength of the other. The armed forces both protect the state and project its will. They are the tool through which the state exerts total control over its territory and its citizens.

Through its state agents, the appointed police and military forces, the state levies taxes on its citizens and holds the exclusive right to condemn to death anyone inside its territory. Benjamin Franklin wrote in 1789 that "in this world nothing can be said to be certain, except death and taxes", both of which are allegedly controlled by the state. Should another actor challenge this privilege and kill a person, or attempt to illegally levy or evade taxes, the state can retaliate through its agents by holding the perpetrator legally accountable and punishing him through any means of its choice, including incarceration and death. The means to implement the law are controlled by the state, although it is noteworthy that the state's use of torture was progressively outlawed in Europe in the eighteenth and nineteenth century as modern society is defined by its political civility which does not condone this method of punishment. Tilly explains that "a tendency to monopolize the means of violence makes a government's claim to provide protections ... more credible and more difficult to resist". The state's promise to protect its citizens from violent crime and foreign attacks is the raison d'être of the state. Bobbitt suggests that in failing to uphold this mandate, the city-state "would have ceased to fulfil its most basic reason for being". The modern state has sought this monopoly and is subsequently defined by sociologist Max Weber in terms of its ability to hold a "monopoly of the legitimate use of physical force within a given territory" which is contingent upon its control over and the effectiveness of its army and police force.

Soldiers make better citizens (and vice versa)

In their rise to power, warlords and rulers have relied heavily on paid mercenaries to subdue the local population, fight off invaders and conquer new lands. This paradigm shifted, however, as sovereigns recognised the fiscal advantages of employing citizens as soldiers. Wars had become forbiddingly expensive in the early seventeenth century as gunpowder changed both the scale and the price of war and the state had successfully disarmed its citizens, thus removing the immediate threat of uprisings. The "sheer cost of warfare overwhelmed the financial resources of all but the most commercialised states" (Tilly 1992: 101) who were forced to develop alternate means of military power and shape their socio-political structure accordingly. It was quickly established that domestic actors were cheaper than hiring mercenaries, especially on a massive scale, and governments began to finance professional armies of trained citizens who could reinforce the legitimacy of the state and project its will domestically and abroad.

Citizens were not only cheaper, they also were perceived to be better combatants, more controllable and more trustworthy than mercenaries – although as explained in Chapters 2 and 3, discipline, training and punishment determine the reliability of a combatant more than alleged character differences. Machiavelli warned his Prince that "if anyone supports his state by the arms of mercenaries, he will never stand firm or sure" due to the unreliable and fickle nature of these combatants. More importantly, however, rulers recognised that they could exert better control over their citizens and their armies by integrating the former into the armed forces where they could be moulded and indoctrinated with the norms that would best safeguard the interests of the state. Plato explained that war "transformed the city from a political unit into a political body: its unity was reinforced by the participation of its citizens and made meaningful only through the citizen's input". Both philosophers place the citizen firmly under the control of the state and at the centre of the state's policies. It is through military service, argues Machiavelli, that the state creates a 'public spirit' among its citizens: i.e. "the recognition on the part of the citizen that his own selfish interests and those of the Commonwealth do, in fact, overlap sufficiently to justify the restraint of the former". For Plato and Machiavelli, war was "a normative exercise because it is only through war that the citizen can become a good soldier, and only a soldier can be a good citizen" (Coker 2010: 145) and therefore submissive to and controllable by the state.

As the armed forces were embedded within society, they were increasingly regarded "not as a part of the royal household, but as the embodiment of the nation" (Howard 2009: 110) and guardians of society. In return, the community is also dogmatised and "[re-]built through the inculcation of these values within the military by education, not institutional devices" (Janowitz in Feaver 2005: 19). The state's preference for citizens over foreigners as state combatants, is inherently normative and reflects the state's deliberate strategy to maintain control over her population by transforming citizens into soldiers in a bid to build a cheap and obedient army. On the other hand, however, an army of citizens increases its control over the government by acting as a check on the powers of the state. This also contributes to society's preference for a citizen-army that reflects the values of the community and protects its interests.

New responsibilities: from protector to provider

The state was never fully insured against popular uprisings, however, even coming from an unarmed citizenry. In their quest to raise taxes and finance their armies, sovereigns have had to make a number of concessions. This

> bargaining took many (...) acceptable forms: pleading with parliaments, buying off city officials with tax exceptions, confirming guild privileges in return for loans or fees, regularising the assessment and collection of taxes against the guarantee of their more willing payment, and so on.
>
> (Tilly 1992: 101)

The state therefore forfeited some of its authority and extended its duties towards its citizens in exchange for their cooperation in paying taxes and serving in the armed forces. On the other hand, the state also redistributed its responsibility of protection to its citizens who were expected to contribute to their safety through private initiatives such as neighbourhood watches, private security and gated communities.

This process progressively led to the development of property rights and democratic representation that the government conceded to the main contributors to its coffers and, much later, to the entire male (and even later female) population. By delegating some power to the citizenry, the state paved the way for the democratic control of the government by the population. Rousseau argues that democratic control is really at the heart of the social contract establishing the legitimacy of the state in the eyes of its citizens. The state was also forced to invest in basic infrastructure such as roads and administrative organisations in order to collect taxes and exercise control over the population throughout the entire territory. Mabee explains that the

> 'security state' (...) represents a situation where the increased penetration of the state into civil society provided the basis for not only more coordination of society by the state, but the reciprocal effect of increased rights and expectations of the citizens of states.
>
> (Mabee 2009: 26)

Sociologist Eugen Weber further analyses this process in *Peasants into Frenchmen*, which describes the generation-long efforts that the French government undertook in the nineteenth century to homogenise and modernise the country into one 'nation' that it could better tax and control. In 1840, France was a cluster of culturally and ethnically heterogeneous villages, where superstitious peasants spoke rural dialects and felt neither unity to the country nor allegiance to the state. Under the Second Empire and the Third Republic, the government developed a nationalisation project to eliminate provincial loyalties and homogenise the citizens under the control of the state. This included the development of new means of transportation (for example through the increased use of the bicycle and the building of roads) which improved communication and tax collection between the periphery and the centre. The standardisation of education and the enforced use of French as the only acceptable language in all the schools and for all administrative processes throughout the country were crucial to the industrialisation and urbanisation of French society. It was through education that the government was able to instil a common sense of patriotism and moral values among its population. Education and military service went hand in hand as the state gained access to all young males and found that it was in its best interest to provide other services: "eventually the health and education of all young males, which affected their military effectiveness, became governmental concerns" (Tilly 1992: 105). The difficulty of the process of turning "peasants into Frenchmen" can be appreciated in view of the many peasant revolts – or *jacqueries* as they are known in France – that shook the country throughout its history. The last *jacquerie* took place in the Vendée in 1793 and to a large extent contradicted the popular philosophy of the levée en masse. Bobbitt explains that the French Revolution transformed France from a kingly state to a nation-state: as the responsibilities of and expectation from the state increased, the nation-state became "not responsible to the nation, rather it was responsible for the nation".

France's modernisation and homogenisation programme was swiftly followed as other states replicated and adapted this strategy, imposing a common language, religion, legal system and building communication and transportation networks. Progressively, the state took on a new role as the provider of public services as well as the guarantor of security. In the nation-state, "most individuals' security is provided by the state – from protection from the internal and external threat of violence to the provision of basic needs – and is therefore contingent on political relationships" (Mabee 2009: 13) such as that between the state, its army and its citizens.

The new security environment

The state has claimed a legal monopoly over the legitimate use of violence ever since its position was consolidated in the 1648 Treaty of Westphalia. By agreeing to a norm of non-intervention and recognising each other's right to exclusive military authority within specific territories, the great European powers set a precedence that they hoped would address their main security threat, which they perceived as coming from other states. This state-centric world endured until the fall of the Iron Curtain, at which point the end of the bipolar system, globalisation and the rapidly changing security environment forced the state to adapt its military apparatus to survive in a new world of non-state threats and actors.

Globalisation and security

By the twenty-first century, the security environment had dramatically shifted from a system of external threats to state security mostly in the form of military threats from other states to threats arising from domestic and transnational actors. The bipolar system and its obsession with the nuclear threat and mutually assured destruction (MAD) kept the focus on the state at a time where globalisation and interdependence were already creating new security threats. Although these security concerns were not ignored entirely by individual states or the international community, the context of the Cold War determined the way in which these issues were viewed and prioritised. The end of the Cold War opened up a space in which security could be re-evaluated in non-state-centric terms. Environmental degradation, the rise of ethno-political nationalism, economic vulnerabilities, diseases, terrorism, organised crime and 'rogue states' highlighted the growing impotence of the state-nation as a solo actor faced with these multifaceted threats.

The term globalisation refers here to the rapidly increasing interdependence of states and non-state actors on a global scale. This interdependence is the result of the invention of new technologies that have made communication and transportation easily and affordably accessible. This has facilitated financial transactions and cultural and ideological exchanges that have tied the economic prosperity and political survival of individual states to that of the international community. The availability of new technologies has created new opportunities, but it has also allowed hostile non-state actors to prosper and therefore expanded the scope of security threats. Hostile non-state actors are defined in this context as individuals or groups that are

> (1) willing and capable to use violence for pursuing their objectives; and (2) not integrated into formalised state institutions such as regular armies,

presidential guards, police or Special Forces. They therefore (3) possess a certain degree of autonomy with regards to politics, military operations, resources and infrastructure.

(Schneckener *et al.* 2007)

Unaffiliated, autonomous and armed, hostile non-state actors are a threat to the Weberian concept of the state because they undermine the state's ability to control the level of violence inside their territory and therefore challenge the state's claim to a monopoly over the use of violence.

Terrorists, insurgents, traffickers and criminal organisations are now able to operate across borders using communication and military technology that were previously accessible only to the state and its agents. They have also gained unparalleled access to weapons, challenging the state's ability to keep its population unarmed; according to a 2007 Small Arms Survey,

> civilians own approximately 650 million of the total 875 million combined civilian, law enforcement, and military firearms in the world, but 'only' 1 million, less than 1%, of small arms and light weapons, are in the hands of insurgent groups.

Furthermore, the ease with which non-state actors transcend national borders has turned local security issues into global problems as threats are no longer isolated within the state nor limited to the periphery.

States have responded to this new framework by creating an integrated network of organisations such as the UN and NATO through which they exercise power and tackle security problems together. Ian Clark explains that

> in so far that this is the case, we are witnessing a diminution of the "go-it-alone" mentality that has been the distinctive hallmark of national security in the recent historical epoch, and a corresponding shift towards what has been called the "transnationalization of legitimate violence".

(Clark 1988: 99)

International security cooperation reflects the recognition that threats are globalised and require a response that goes beyond the state's individual abilities. This has contributed to establishing a norm by which all international actions require an official stamp of legitimacy from the international community of states.

Globalisation has transformed the way in which security is perceived and handled, with the result that the providers of security are also changing. Non-state transborder threats require responses that are not necessarily achievable through state-centric military might. States, in their pursuit of peace and security, have become increasingly integrated into the international community and have begun to look for alternative means with which to face the next challenge. This has led to a shift of the traditional role of the state to enable it to face the new security environment while maintaining power and control.

The state as facilitator

Philip Bobbitt argues that the state is continuously evolving and that its future lies not in its former role as the provider of welfare but in a role as facilitator or regulator of services. The state has indeed undergone substantial transformation since the fall of the Iron Curtain. It no longer acts alone to provide security and public services but has used international institutions and private providers as vehicles to enact its policies. This is an inevitable result of the growing costs of public services and public debt and reflects the philosophy of privatisation that has swept across most of the developed world.

Health care and education, two pillars of the nation-state, have already undergone privatisation in certain countries. The telecommunications and transportation industries, ranging from postal services to roads, railways and phones have also been merged into privately held companies that are arguably more cost-efficient, if less democratic, than the state. The profits derived from these services are in turn taxed by the state that maintains control over the licensing and profitability of companies established or operating inside the country. The privatisation of public services has been controversial, with critics accusing the state of failing to uphold its responsibility towards its citizens as some services have become selectively unavailable except to those who can afford them.

Security and control over the use of violence, a guarded feature of the state up to recently, are also being outsourced and privatised as states seek to reduce costs and keep up with the changing security environment. For example prisons, a fortress of state control over the population, are increasingly run by private companies that are charged with the welfare and disciplining of the state's felons: In the United States, 7 per cent of all prisons are owned and run by companies, as are 9 per cent of prisons in the UK and 18 per cent of prisons in Australia. Israel's Supreme Court, on the other hand, ruled that privately held prisons were unconstitutional as the state is considered to be the only entity with the right to exert force and incarcerate its citizens and residents.

Bobbitt suggests that the role of the state has shifted from "nation state to market state". The nation-state, as described above as the 'provider', offered "free public mass education, universal suffrage, and social redistributive taxation" but above all it "sought its legitimacy in the betterment of the welfare of its people". The market state, he conceptualises, "promises instead to maximize the opportunity of the people and thus tends to privatize many state activities and to make voting and representative government less influential and more responsive to the market". The state is "no more than a minimal provider or redistributor" in this international marketplace. Bobbitt argues that, by adapting its function and responsibilities and featuring its role as a facilitator promising to maximise the total wealth of the society, the market state ensures its own survival as the primary actor in international relations. Stephen Krasner and Anna Leander further this argument by explaining that the state never actually held a monopoly over the use of force, and that it is not legitimised by its control over the armed forces but by its proven ability to provide security and social benefits to its citizens, either directly or by delegating these tasks with the objective of optimising economic opportunities.

In view of the rapid demilitarisation of Western societies since the end of the Cold War, the role of the state is being transformed: conscript armies and civic duty are replaced by smaller and more professional armies of volunteers. Elke Krahmann explains that the "non-existential nature of most contemporary threats decreases citizens' willingness to accept large military budgets and to contribute personally to national defence" (Krahmann 2010: 247). Martin Shaw calls this the 'post-military' societies, in which the mutual contract between the state and the individual is broken: as citizens no longer desire nor have any obligations towards the state, neither does the state need to act as the sole provider of security and can instead base its legitimacy on its role as a facilitator of security and economic progress.

Changing norms on combatants

The shifting role of the state has also affected the armed forces. 'Postmodern' militaries, facing neither existential threats nor the risk of invasion, have undergone major organisational changes and experienced a loosening of ties with the nation-state. Moskos claims that the "sense of identity with and loyalty to the nation-state is 'decomposed' in Postmodern society", paving the way for non-nationals to find a place in the national armed forces (Moskos *et al.* 2000: 4).

Although the state has claimed a preference for its own citizens to serve in the armed forces, the public apathy and even hostility towards civic duty that has prevailed in 'post-military' societies have compelled many Western governments to change recruitment laws in order to staff their armies. These laws have been enacted by the state, and despite drawing in foreigners and foreign private companies into the defence system of the state, have been largely unopposed by the country's citizens who, ultimately, do not want to volunteer for military service. Indeed, Chapter 3 shows that the ""postmodern motivation' characterizes soldiers who enter the military more for the desire to have a meaningful personal experience than out of either national patriotism or an occupational incentive" (ibid.: 6). Consequently, armies are increasingly staffed with foreign nationals, reminiscing of a time when the state did not fully control its population through enforced military service. Hedley Bull refers to this trend as 'neo-medievalism'.

In the United Kingdom, shortfall in recruitment numbers have forced the government to remove restrictions requiring citizens from Commonwealth nations to live in the United Kingdom for at least five years before applying to join the British armed forces. Between 1998 and 2008, the number of soldiers from the Commonwealth enlisting under the Union Jack leapt from 200 to 6,600. Official figures include 2,000 Fijians, 975 Jamaicans, 720 South Africans and 1,000 Zimbabweans and Ghanaians among the other 58 nationalities represented in the British Army (*Guardian* 5/4/2008). The recruitment gap therefore is being met by Commonwealth soldiers, with one in five soldiers expected to be non-British by 2020. This has raised some concern among the armed forces and the population who object to this 'invasion' of foreigners, lured by the promise of British nationality and public service. Sir Richard Dannatt, the former Chief of the General Staff, proposed that foreign soldiers "be capped at 10 percent of the total strength to protect the 'Britishness' of the army" (*Guardian* 2/2/2007). The Ministry of Defence, however, recalled Britain's "long and successful tradition of employing and integrating overseas personnel" to justify the continued use of this practice to the population. The British armed forces, therefore, not unlike the French Army, is being increasingly hybridised by the integration of foreigners into the defence system of the country. Ironically, the British Army is also cutting down its numbers drastically in its programme to reform the armed forces, increasing its future dependence on military and security contractors.

The United States has experienced a similar situation with the narrowing of its recruitment base and its military reforms. Author Darren Moore calls the separation between civilians and their army 'the Great Divorce'. The Afghanistan and Iraq wars forced President Bush to change recruitment and incentive policies by offering all non-citizens on duty eligibility for immediate US citizenship in order to encourage more people to enlist in the US Army. In 2003, approximately 3 per cent or 37,000 active-duty soldiers were non-US citizens, with an additional 13,000 foreign reservists. The Chief of the Army Reserve, Lieutenant James Helmly, defends this policy by claiming that "we must consider the point at which we confuse 'volunteer to become an American soldier' with 'mercenary'". Although Lieutenant Helmly sees a significant difference between the two, by offering citizenship as a currency for enlisting in the army, however, the US is effectively 'buying' its volunteers. These men are motivated to fight, not out of patriotism, but for the promise of a new life with relative economic and political security. Nonetheless, soldiers, whether foreign or domestic, as long as they are integrated into the armed forces, are under the authority of the military legal system and are consequently accountable to and controllable by the state. This legal and political accountability marks a significant distinction between the two categories of soldiers and mercenaries and makes one legitimate in the eyes of the population while the second remains in the shadows of the law.

Another characteristic of postmodern societies is the increased dependence of the military on the private sector, highlighting a new permeability between civilian and military structures. Chapter 6 describes the war in Iraq as a platform for unprecedented initiatives in outsourcing and privatising military services to domestic and foreign private military and security companies. The United States has invested over US$75 billion in private military support according to the Congressional Budget Office. By comparison, the UK MOD spent approximately half of its £34 billion defence budget on purchases and services from the private sector in 2006. The state's success at changing its military composition unopposed by the public suggests that norms regarding the nature of the combatant are set by and entrusted to the state. Although the word 'mercenary' remains controversial, foreigners and private contractors have nonetheless become more or less forcefully and successfully integrated into the defence structures of the nation-state (France) and the market-state (the United States).

Authority and control

This suggests a historical change in norms regarding the concept of the state: it no longer holds a monopoly of violence but a monopoly of security. The state maintains its legitimacy in the eyes of its citizens by enabling security to reach the population through mediums of both the public and private sectors. The post-modern or market state provides "insurance against contingency" (Giddens in Booth 1998) and remains ultimately, but not directly, responsible for the security of its territory and its citizens. The state retains the exclusive monopoly to delegate functions to other actors but may also penalise these agents if they challenge their instructions or fail to fulfil their mandate.

In *Politics as a Vocation*, Weber explains that "the right to use physical force is ascribed to other institutions or to individuals only to the extent to which the state permits it". By this token, the state remains the "ultimate arbiter of the legitimate uses of force" (Ortiz 2010: 5) even when it assigns security and defence functions to private and foreign actors. Thomson and Leander stress that it is important to distinguish between control and authority as the state may outsource its security without losing control over the use of violence inside its territory.[1] This was illustrated in Chapter 5 where it was shown that President dos Santos' authority in Angola and control over the exercise of violence were not weakened by his contract with a foreign private military company. It is, however, also a function of the nature of the state, which can be profoundly different in different developing countries.

Bobbitt suggests three side-effects of the market state that reflect this change in norms and expectations vis-à-vis the state: first, the market state "will require more centralized authority for government, but all governments will be weaker" from outsourcing and privatising their responsibilities. This leads to a devolution or complete loss of authority as institutions and private actors take on these duties. Second, although "there will be more public participation in government, it will count for less, and thus the role of the citizen qua citizen will greatly diminish and the role of the citizen as spectator will increase". Finally, the welfare state will be reduced while "infrastructure, security, epidemiological surveillance and environmental protection (...) will be promoted by the State as never before" (Bobbitt 2003).

Increased surveillance and security is necessary, as the state's commitment to its citizens and relationship with violence is one of comparative value, according to Douglass North. The state defines itself in terms of a superior ability to exert violence within its territory, not in terms of a monopoly over the legitimate use of force or as a direct provider of security: "the state is an organisation with a comparative advantage in violence on a territory whose boundaries are determined by its power to raise taxes" (North 1981: 21). Whereas the state may privatise or outsource security, it must maintain the ability to reverse its policies by exercising more power than any agent, private or public that operates within the boundaries of the state, and bringing these agents to heal whenever necessary. Indeed, nearly every country in the world invests in its own military force and defence system, even while it promotes outsourcing as a strategic solution to security problems and foreign policy ambitions. Mercenaries and private military companies, however, have shown at times an incredible disregard for national and international laws, and yet have remained beyond the legal arm or political will of hiring states. The lack of accountability of these actors, therefore, suggests a weakness in the state's authority or in its desire to control the actions of its hired agents.

Accountability

Accountability is a fundamental characteristic of the state on two levels: (1) it defines the state's ability to exert control over its agents and therefore establishes it as a legitimate sovereign actor, and (2) it upholds the social contract between the state and its population. The ability or lack thereof, to hold non-state actors accountable, is the key problematique that determines the social norm regarding the state's official use of foreigners, mercenaries, and private military and security companies. Mercenaries are considered to be dangerous because they do not represent the values of the community for which they are fighting, they are unrestrained in their actions, and they are untrustworthy. Despite international, domestic, and corporate efforts to legislate, limit and punish the use of mercenaries and contractors, these non-state actors have continued to operate in a largely unrestricted and unaccountable environment therefore challenging the authority and political will of the state and undermining the democratic institutions of the nation.

International laws

Mercenaries are considered to be illegal combatants according to the norms of international law that were drafted by the community of states. The laws of war outlined in the 1949 Geneva Convention and the 1977 Additional Protocol I and II preclude mercenaries from being considered legal combatants. This removes them from the legal protections that are enjoyed by soldiers and civilian combatants during a conflict, in particular the right to being treated as a prisoner of war, as outlined in Geneva III which entitles the prisoner to medical care, humane treatment and protects him from torture and murder at the hands of his captors. Conversely, since mercenaries do not fall under the prescribed laws of war, neither are they bound by accepted norms of behaviours that limit the actions of legal combatants.

The 1972 (adopted in 1977) Organisation of African Unity (OAU) Convention for the Elimination of Mercenaries in Africa clearly describes mercenaries as a threat to "the independence, sovereignty, territorial integrity and the harmonious development of Member States of the OAU". The OAU Convention commits its member states to taking legislative measures to fight the recruitment, training, equipping, financing and use of mercenaries within their territories in accordance with Article 3 (f) and to enforce these laws with "severe penalties for offences" in Article 4.

In 1989, the United Nations General Assembly adopted the International Convention against the Recruitment, Use, Financing and Training of Mercenaries which further advanced the campaign against mercenarism. The UN Convention includes all instances of recruitment, use, financing and training of mercenaries as offences, and signatory states engage to cooperate in the prevention of these crimes and punish offenders by "appropriate penalties" as described in Articles 5 and 6. Furthermore, signatories are expected to share any information on mercenaries with other affected parties (Article 7) and to abide by mutual judicial assistance when applicable (Article 13). The 1989 Convention is signed by 17 states but needs 22 signatures in order to become law. Significantly, France, the United States and the United Kingdom are not party to this convention, and Angola has signed but not ratified the treaty.

Despite the aforementioned conventions and other resolutions passed by the General Assembly, "there is no total ban on the use of mercenaries under international customary law" (Kinsey 2008). Efforts to curtail the activities of mercenaries have been largely ineffective in international law and each of these conventions suffers from a lack of enforcement provisions. Furthermore, there are "no international instruments in existence which could regulate corporations (or mercenaries) on the international stage (and) there are no courts competent to adjudicate on corporations (and PMCs) in the international context" (Sheehy *et al.* 2008: 53). Without enforcement, legal conventions act only as political statements reflecting general norms. Finally, signatory states have been reluctant to enact domestic legislation to enforce these conventions, and some have even transgressed the treaties altogether, as exemplified by Angola's contracting with the PMC Executive Outcomes in 1993. Overall, it appears that there is a "lack of political will to achieve a global regulatory regime" (ibid.: 53) to deal with the perceived problem of mercenarism.

Domestic laws

Whereas international law has failed to hold mercenaries accountable, domestic legislation has been selectively pro-active at regulating the activities of non-state actors. Some states have outright outlawed the recruitment and use of mercenaries within their territorial jurisdiction: in Australia, "the recruitment of mercenaries and the fighting of Australians in non-governmental forces abroad is an offense"; Denmark has prohibited "the recruitment for foreign forces and the participation of Danish citizens in foreign armed groups"; Italy and Russia have forbidden "mercenary activities and the recruitment, training, financing and use of mercenaries" (Jäger and Kümmel 2007: 409).

Although France has not signed the 1989 Convention, since 2003, it has criminalised mercenary activities which are punishable according to its Penal Code by a fine of up to €75,000 and potential imprisonment (L436–1 to 5). The law specifically applies to mercenary activities that threaten the territorial integrity of another country, and does not prevent French citizens from volunteering in foreign forces. The UK, with the Foreign Enlistment Act of 1870, has made it unlawful for British citizens to join the armed forces of any nation at war with Britain or its allies. Like France, however, Britain does not outlaw the recruitment or use of foreigners in the armed forces; Both the French Foreign Legion and the Gurkha regiments are exempt from the 1949 Convention's definition of mercenaries by virtue of their formal integration into the national armed forces and accountability to the state. Although a Green Paper suggesting options for regulating the private military industry was published in 2002, no action was ever pursued, and the 1870 Act remains the only legislation covering mercenary activities in the UK – although "there have been no successful prosecutions under this act" (Sheehy *et al.* 2008: 138).

South Africa is often acclaimed as the forerunner of legislation that regulates mercenary activities. The post-apartheid government sought to control the "embarrassing" availability of mercenaries who were being recruited in South Africa and participating in violent conflicts worldwide. South Africa enacted the Regulation of Foreign Military Assistance (FMA) Act in 1998 which "precludes any South African citizen from participating in armed conflict, nationally or internationally, except as provided for in terms of the Constitution or national legislation". The legislation therefore established a licence-based regulatory scheme that entrusted the National Conventional Arms Control Committee (NCACC) with authorising specific contracts on the provision of security services abroad. Notably, on 24 October 1997, the South African PMC Executive Outcomes was awarded a "government-issued licence to continue with its work" (Barlow 2008: 518). In 2007, in the wake of the failed 2004 mercenary coup in Equatorial Guinea involving a majority of South Africans, a second bill on the Prohibition of Mercenary Activities and Regulation of Certain Activities in Country of Armed Conflict was signed by President Thabo Mbeki. To this day, the bill has not been enforced and South Africa continues to be a source of contractors to conflicts in Africa and the Middle East. Mark Thatcher and the mercenaries arrested in Zimbabwe and repatriated to South Africa were the first men to be prosecuted under South Africa's mercenary laws. According to mercenary pilot Niel Steyl, if not for their plea bargain, the government would have been incapable of successfully prosecuting these men due to loopholes in the legislation and a lack of evidence.[2] Despite its best intentions, the FMA and the latest mercenary bill remain 'toothless' threats.

The regulation of mercenaries and private military and security companies in the US falls under the International Traffic in Arms Regulations (ITAR) that came into force in 1998. ITAR establishes a licensing system which is overseen by the State Department's Office of Defense Trade Controls. One of the principle criticisms directed against ITAR is that, under the licensing regime, "Congress does not need to be informed of a contract in advance of the issuance of a related license unless the contract is valued as $50 million", which does not happen very often because most contracts are valued less or can be split into partial subcontracts. This stipulation consequently circumvents parliamentary oversight. Nonetheless, the US model of regulation arguably offers "some control of PMCs that are operating in weak states unable to exercise any effective control" by "setting (minimum) standards for PMCs using the US as their home base" through the licensing conditions required in the US (Jäger and Kümmel 2007: 146). Legislation in the United States, however, falls short of holding companies or mercenaries accountable for offences committed abroad. US criminal law is not enforceable outside of the territory, and the Uniform Code of Military Justice only applies to military personnel. Furthermore, the 2000 Military Extraterritorial Jurisdiction Act "extends the realm of the Code to civilians contracted by the Pentagon, but does not cover transgressions committed by civilian contractors working for other government agencies or for foreign clients" (Jäger and Kümmel 2007: 146). Currently, not a single employee or private military and security company has been found culpable of a crime, despite the many incidents in Iraq and Afghanistan.

The global nature of PMCs makes them difficult to regulate, monitor and prosecute in the first place. National legislation is limited to territorial jurisdiction, and there are no mechanisms in place to exercise oversight on the behaviour of the PMCs abroad. Even "if a state has the legal means for extraterritorial enforcement of PSMCs and their personnel, the company concerned still can avoid prosecution by relocating to another country" (ibid.: 416). Overall, national legislation over the use and recruitment of mercenaries and PMCs has so far been ineffective and half-hearted. This suggests that governments are not seriously concerned with the threat of mercenaries, or that they want to keep open the option of using these non-state resources themselves.

Corporate regulation

Where domestic legislation has failed, private military and security companies have argued that they are naturally regulated by market forces and in addition, they abide by a system of self-regulation. Furthermore, as corporations, PMSCs are also regulated by the contract laws and corporate laws of the country in which they are established. Regardless of these self-justifying allegations, however, the activities of mercenaries and military and security corporations are only loosely regulated, and they are rarely, if ever, held accountable for offences that they may have committed while carrying out a contract.

Tim Spicer, the CEO of Sandline and Aegis, claims that private military providers are regulated by the free market and open competition: "we do not run a mercenary outfit. We provide regulated, professional military assistance to established governments" (Spicer 2000). Assuming that PMSCs are operating in a free market and competing with other companies, their reputation is a fundamental selling tool and therefore an incentive for these companies to abide by rules of conduct and show professionalism, reliability and integrity. Concern for their reputation ensures that companies will hold their employees to high standards in order to avoid public scandals and negative publicity. Isenberg argues that mercenary groups "fear publicity that would result if a mercenary subsidiary carried out a massacre or other egregious human rights abuse. Thus, a corporate association might well be a restraining influence in regard to the battlefield conduct of such groups" (Isenberg 1997). Executive Outcome's emphasis on good civil–military relations stems from the knowledge that "the fastest thing that would get us out of business is human-rights violations" (Barlow 2007). A PMC has to be able to wrap up its mission in order to be able to accept other contracts and show a track record of successful and efficient performance. This also supports their profit-maximising objectives and contradicts allegations that PMCs have an incentive to prolong their contracts by encouraging instability in their theatre of operation. PMCs that encourage a situation of conflict risk losing both their reputation and their future contracts.

Despite allegations that the free market regulates the private security industry, Chapter 6 shows that the system of contracting, at least in the US, and likely elsewhere, has been found to be uncompetitive, with major contracts allocated to companies in a no-bid process. Governments are not following the rules of the free market and rewarding contracts to companies who offer the best services in terms of quality and price. On the contrary, personal relationships and connections repeatedly determine the allotment of security contracts. As a result, "the most successful contractors are not necessarily those doing the best work, but those who have mastered the special skill of selling to Uncle Sam" (*New York Times* 4/2/2007). Recognising this opportunity, "the top 20 service contractors have spent nearly $300 million since 2000 on lobbying and have donated $23 million to political campaigns" (ibid.). Singer explains that privatisation is only successful if

> a contract is competed for on the open market, if the winning firm can specialize on the job and build in redundancies, if the client is able to provide oversight and management to guard its own interests, and if the contactor is properly motivated by the fear of being fired. Forget these simple rules, as the US government often does, and the result is not the best of privatization but the worst of monopolization.
>
> (*Foreign Affairs*, March/April 2005)

In the absence of regulations and to advance their credibility, private military and security companies have adopted self-regulatory codes of conduct under the political umbrella of ISOA – the International Stability Operations Association (formerly IPOA). Members of what is ostensibly a trade association agree to abide by the IPOA Code of Conduct. Although this Code "encourages" all signatories to abide by international humanitarian law, human rights law and other rules of engagement and international treaty, this suggestion is part of the preamble and is therefore non-binding. Members pledge to operate with "integrity, honesty and fairness" in so far that this is "possible and subject to contractual and legal limitations" (Article 2). Furthermore, the Code "is not enforceable in any meaningful way" (Sheehy *et al.* 2008: 115). Sheehy and Maogoto highlight that "there is no complaint mechanism in the Code, which decreases the likelihood that any breach of the Code will be reported. There is also no mention of a compliance office or a regular audit process". Companies who breach this Code face no sanctions other than, possibly, being dismissed from the association.

Nonetheless, most states have a legislative framework to control the actions of private corporations: entities within the territorial jurisdiction of a nation are bound by contract law and corporate law, including provisions for registering a company, internal governance and accountability. Corporations are legal entities that have privileges and liabilities that are distinct from their members. Corporate law is concerned with the regulation of the internal structure of the corporation, and limits the rights, duties and liabilities of the individuals associated with the corporation: directors for instance, "are protected from the consequences of their actions done on behalf of their principles" (ibid.: 44). The logic behind this law is to encourage a certain amount of risk-taking within the company which would enable it to maximise profit. Likewise, shareholders are generally exempt from "liability for the actions or obligations of the corporation and decisions taken in the process of pursuing profit" (ibid.: 45). Directors and shareholders are "mere agents of the corporation and so take no personal liability for the actions executed on behalf of the corporate principals" (Sheehy *et al.* 2008: 46).

Directors, shareholders and corporations are as a result shielded from criminal prosecution and there is currently no legal regime to consistently hold non-state actors and their affiliations and employees accountable for their actions. Public outrage is perhaps the only instance where private military and security companies have been challenged, put under criminal investigation, or lost a contract, as evidenced in Iraq after the Abu Ghraib scandals and the allegations of negligent and even criminal behaviour from individual contractors. In most cases, however, despite being under investigations, select PMCs involved in scandals are still awarded new contracts regardless of any criminal proceedings.

The Montreux Document

Following the series of international law violations and human rights abuses committed by private military companies in Iraq, the international community began the difficult task of drafting and implementing a code of conduct to regulate the actions of these companies. It has now been five years since the release of the *Montreux Document on pertinent international legal obligations and good practices for States related to operations of private military and security companies during armed conflict*, known as the Montreux Document. This non-legally binding document has been signed by 46 states and the European Union and sets about to correct the legal vacuum in which PMCs have been operating.

The Montreux Document, spearheaded by the Swiss government and the ICRC, recalls the international humanitarian and human rights law obligations of states and reminds them of their accountability for the behaviour of PMSC personnel that operate on their behalf during armed conflict. It also describes Good Practices that help support and implement the obligations of states through national measures and sets the stage for an International Code of Conduct for Private Security Service Providers (ICoC).

The Montreux Document and ICoC have had mixed results. A report published by the American University Washington College of Law in December 2013 explores the difficulties of enforcing state commitment vis-à-vis the implementation of regulations. For instance, despite being a signatory to the Montreux Document, "the US has not enacted a comprehensive system of laws and regulations to hold PMSCs and their personnel criminally accountable for violations of national and international law" (Perret 2013: 8). Likewise, five years after the UK government began pursuing a policy of voluntary self-regulation for the private military industry, "this approach (has not been) effective because it leaves the U.K. government with little ability to influence PMSCs with whom it has no contractual relationship" (ibid.: 9) and there are no institutions in place to monitor whether this policy even works. In Iraq, even though the Iraqi government has taken over the security of the country and implemented rules on the vetting of personnel, registering of vehicles, and weapons, audits and licensing, "it (still) does not meet all the criteria laid out in the Montreux Document" (ibid.: 10).

Nonetheless, the Montreux Document is a work in progress. According to Anne-Marie Buzatu[3] at the Geneva Center for the Democratic Control of the Armed Forces (DCAF), private military and security companies are increasingly contacting DCAF for information on and to be included in the International Code of Conduct. By September 2013, membership had grown from 58 to 708 companies from at least 70 countries. Companies become signatories to ICoC, publically affirming their responsibility to respect human rights and humanitarian law. Although the ICoC creates no legal obligations or liabilities, signatory companies are expected to commit to international standards of behaviour and improve oversight and accountability of their personnel. According to Buzatu, companies have an incentive to cooperate with the ICoC as certification sends a signal that these companies are trustworthy and are in compliance with existing international laws.

In addition, the Montreux Document and ICoC have raised awareness and fostered better regulation of the industry at the national level, as evidenced in the *Montreux Five Years On* report. The next step in this initiative is to build external monitoring and implementation mechanisms to ensure that signatory companies follow up on their commitments. Holding companies and their personnel legally responsible for human rights violations, however, is still a distant project due to a lack of commitment and political will both domestically and internationally, and voluntary compliance may be the only viable option to regulate the industry at this time.

Democratic control

One of the foremost problems regarding the use of mercenaries and private military companies is that it has undermined democratic institutions in hiring states by bypassing the influence of the military. The armed forces have not only been a tool for the state but also an element of democratic control limiting the powers of the government. Samuel Finer claims that the military sees itself as the servant of the state, rather than of the government in power, and Samuel Huntington perceives the armed forces as responsible for the state. Because the armed forces are traditionally made up of citizens, the army is imbued with the values of the citizenry who are in turn empowered by being represented inside the military institution. Civil–military integration has "ensured civilian control over the armed forces, and contributed to public knowledge about security and defence policies" (Krahmann 2010: 251).

Michelle Small argues that in Africa, the "privatisation (of security and military functions) dramatically reduces the citizenry's ability to influence governments' policy, diminishing popular sovereignty and inviting the public to wonder whether decisions to militarily intervene are influenced by PMFs who supply the weaponry and services" (Small 2006: 19). The same is applicable in other sovereign states. As was previously discussed, the privatisation of force has occurred without parliamentary oversight and governmental accountability.

Table 7.1 Comparative study of non-state warriors

	French Foreign Legion	*Executive outcomes*	*American contractors*
Nationality	International	South African	American
Sector	State-controlled, integrated into Armée de Terre	Private	Private
Hiring agent	French Army	MPLA government	US government and private sector
Duration	Indefinite	3 years	Short-term contracts
Function	Combat and peacekeeping	Training, advising, occasional combat	Defensive/ protective, occasional combat
Legal accountability	French law	Independent/*ad hoc*	Non-binding regulations
Popularity with regular army	Accepted but marginalised	Initial resistance then cooperation	Reluctant collaboration, resentment
Popularity with population	Celebrated by French population	General indifference	Unpopular both with Iraqi and US population
Impact	History of successful military campaigns	Successful. Achieved military victories but no defeat of enemy	Inconsistent. Successful protective services but strategically harmful to overall campaign

Krahmann explains that the outsourcing of military functions has had a "negative impact on public information and parliamentary control over defence spending and the use of armed force in deployed missions" (Krahmann 2010: 249). This was illustrated in Chapters 5 and 6 where the privatisation of military services has expanded the government's autonomy with regard to their foreign policy and the selected use of force, as in the cases of Angola and the US government in Iraq. PMCs have acted as a force multiplier, but have also given the government an opportunity to act independently of public opinion and without regards for public support. Of course, this may be to the advantage of a government who wishes to carry out an unpopular or unsupported policy domestically or externally.

Private corporations are protected from the checks and balances of governments and are not required to be transparent and accountable. Furthermore, they are not burdened by political responsibility nor do they owe allegiance to the population. The freedom of private military and security companies to operate without legislative or legal oversight removes them from "the public domain of accountability, legitimacy and control" (Small 2006: 4). These non-state actors may act independently, carry out a government contract blindly and unquestioningly, and can count on the legal vacuum to escape from being held accountable for their actions. Because PMCs have the ability to inflict unrestricted violence, they undermine the democratic control of the armed forces and of the government, weakening the social contract between the military, the government and the population.

Conclusion: state control in the new century

Norms are changing: the rise of contractors and mercenaries is eroding the state's control over its population and over the combat operations of its agents. The impact of these non-state actors varies according to the state that employs them and to the conflict in which they are involved. Overall however, the lack of accountability that these men enjoy has affected the state's credibility in the eyes of its citizens and of its armed forces.

International and domestic laws fail to hold private contractors, military corporations and mercenaries accountable for their actions on the ground. The state's control over its territory is challenged by the presence of non-state actors who are beyond the legal control of the state. The legal vacuum surrounding non-state actors is arguably a deliberate strategy that the state employs in order to expand its military reach beyond the wishes of its population. The lack of public reaction however, despite the scandals created by African mercenaries, the Foreign Legion in Algeria and contractors in Iraq, shows that the populace remains largely unaffected by the choice of combatant – as long as they are free to ignore their own civic duty to the state and to each other. Ultimately, the population does not care about the norm on combatants, giving the state the possibility of setting and changing the standard.

Nonetheless, the hybridisation of the armed forces has affected the national armies of the states in question. The military has been a symbol of nationhood and state control for the past 200 years. The integration of foreigners, although they remain under the juridical system of the armed forces, challenges the national identity and emotional symbolism of the army as a representative of the people. The lack of accountability of these non-state actors and the frequent differential in pay also serve to demoralise the state's soldiers, discourage enlistment and subsequently weakens the army. This in turn undermines the military institution as an effective check on the powers and behaviour of the government.

The use of private contractors, military corporations and mercenaries to serve the government's ambitions may appear to give the state powers and capabilities in excess of their material endowment. The caveat, however, lies in the potential loss of control that the state may experience. Without a strong and loyal army to implement the will of the state and punish criminals and perpetrators of illegitimate violence, the state is a mere puppet in the hands of corporations who remain unaccountable and beyond state control.

Notes

1 Thomson, *Mercenaries, Pirates, and Sovereigns: State-Building and Extraterritorial Violence in Early Modern Europe*, 223.; Leander, Eroding State Authority? Private Military Companies and the Legitimate Use of Force.
2 Interview with Niel Steyl, Johannesburg, May 2011.
3 Interview with Anne-Marie Buzatu, Geneva, June 2013.

Conclusion

Although mercenaries, contractors and legionnaires continue to excite the imagination of the public, they remain uncomfortable symbols of violence. They are feared for their lack of accountability and their foreignness to a conflict in which it is assumed they have no intrinsic interests. If they are not trusted by the population on whose behalf they are hired, then should they be considered at all as providers of national security? This book has considered three aspects of this question: first, it has investigated how non-state warriors compare to citizen-soldiers on the battlefield. Second, it has assessed the relationship between citizen-soldiers and non-state warriors when these two combatants are forced to work together in the battlefield. And finally, it has evaluated the impact of hybridisation on the national security of select states.

Contesting military values

The armed forces, as the principle guarantor of national security, are an integral part of the nation-state: They protect the integrity of the territory from domestic and foreign threats; they defend the authority of the state, and they promote the security of the citizens. The army has traditionally been the instrument through which sovereigns have built up their power and controlled their population. It has also been the safeguard of democratic institutions and a check on the ambitions of the government, as explained in Chapter 7. The national army is a vital institution that has traditionally enjoyed preferential treatment both in terms of its budget allocation and through its glorification within the state. Soldiers have subsequently developed a strong sense of their standing within society, and have nurtured military traditions that embody the nation's shared values but also reflect their own uniqueness and importance to the state.

The army's identity, influence and role within the state vary along with the culture, history and traditions of their countries. Consequently the exposure to and integration of foreign and private combatants with the national armed forces have affected armies in different ways. The chapter on Executive Outcomes, for example, revealed that the privatisation of the military sector can be a successful foreign policy tool for weak states that lack democratic legitimacy and an effective military arm. In Angola and in Sierra Leone, where the armed forces were in dire need of training, equipment and discipline, Executive Outcomes arguably played a decisive role in the resolution of the civil wars. On the other hand, states that have a strong civil society and a relatively powerful army can be more susceptible to the hiring of a foreign company that has the potential of sidelining the national army. This was demonstrated in Papua New Guinea in 1997, when the National Defence Forces successfully staged a military coup against the Prime Minister after he had signed a US$36 million security contract with the British PSC Sandline. The potential effectiveness of hiring a private military company, therefore, cannot be assessed without taking into consideration the strength of the army's institutional identity.

Contractors and mercenaries appeal to states because they are a short-term, monetarily advantageous force-multiplier that has the potential of increasing the military power of a relatively weaker state against its enemies. The impact that these non-state warriors have on the military institution, however, is not negligible: contractors bypass the ranking and compensation system of the armed forces, fall outside the command and control of the military hierarchy, and are unaccountable for their actions. These non-state agents are in direct competition with the armed forces, monetarily and politically. The privatisation of security may also encourage governments to neglect their own military institutions in favour of private corporations that can allegedly do the same thing at a lower price. The integration of foreign non-state combatants into a state's military policy inevitably weakens the influence and the social and ideological cohesion of the national army. This was demonstrated in Chapter 5 which highlighted the tensions that characterise relations between US soldiers and private contractors in Iraq. The privatisation and hybridisation of the armed forces is a potentially dangerous experiment that runs the risk of devaluing the military's standing in society. It demoralises the armed forces and erodes its unique responsibility for and commitment to the security of the state. The national army is to a large extent a symbol of sovereignty, democracy and citizenship.

On the other hand, the integration of foreigners into the national armed forces has been as contentious, albeit relatively more successful, than the outsourcing of military services. Foreign combatants, who have similar criteria to mercenaries, considering their foreignness to the conflict, have been crucial in advancing French foreign policy. Their success in battle and celebrity both in France and abroad testify that nationality is not a prerequisite for a good combatant, nor are national pride and security necessary to motivate a warrior into battle; *esprit de corps*, training and discipline can be just as effective.

Si vis pacem, para bellum[1]

History has shown that conflict is intrinsic to mankind. Philip Bobbitt explains that

> war is not a pathology that, with proper hygiene and treatment, can be wholly prevented. War is a natural condition of the State, which was organized in

order to be an effective instrument of violence on behalf of society. Wars are like deaths, which, while they can be postponed, will come when they will come and cannot be finally avoided.

(Bobbitt 2002: 819)[2]

Sixty seven years without a major war in the Western world have changed attitudes towards security. Despite twenty-first-century threats ranging from terrorism to environmental meltdown, there is no longer a perceived immediate existential threat to Western civilisations. This new security environment has forced a re-evaluation of risk from both the state and its citizens, leading to the reorganisation of the armed forces, with military security being selectively delegated to non-state actors.

The recent shift away from conventional warfare has decreased the immediate need for large standing armies and encouraged a change in defence mechanism from conscript to professional armies. Recent fiscal difficulties have also forced many Western states to downsize their defence budgets and review their military objectives. Salaries and pensions have been cut, reflecting a change of attitude towards the men and women who fight for their country. Indeed, Western societies have recently witnessed a demotion of military values, as fewer citizens are willing to sacrifice their comforts for the security of their state. Contemporary attitudes towards risk, sacrifice and responsibility are inevitably weakening the nation's commitment and capability to provide for its own security. Consequently hybridisation has emerged as an attractive option, allowing states to cut their military expenses without compromising the political will of their governments.

Most states have favoured private military companies of Western origin. These are perceived to be more trustworthy and culturally compatible with the military and political objectives of their employer. With fewer citizens in the West exposed to military training, however, third-country nationals are already making up more than half of all contractor personnel in Iraq. As the West moves away from conscription, former soldiers from Colombia, Pakistan, China and India, among others, will increasingly make up the bulk of contractors hired by Western companies and governments. These mercenaries have a very different military tradition from the armed forces of Europe and the United States, and their integration in or cooperation with the national armies may prove to be particularly difficult. On the other hand, China is exclusively using Chinese private military companies in her ventures in African states.

An over-reliance on private military companies has not only weakened the effectiveness of the armed forces, it can equally jeopardise the stability and security of the state. States that outsource their security needs to private military and security companies also risk exposing themselves to the organisational errors and limitations of private corporations. This was aptly illustrated in the run-up to the 2012 Olympic Games as G4S, the company entrusted with the security of the Olympic venues, failed to deliver the necessary personnel to fulfil their contract. This prompted the British Army to "come to the rescue" with a peacetime deployment of over 18,000 soldiers. Although the British Army was able to expertly handle the situation, this example is one of many that show the continued necessity for a state-controlled army, whether this is made up of citizens, foreigners, or both.

There are no guarantees that the future will be peaceful, and all indications suggest otherwise. The erosion of civic-militarism and the demilitarisation of societies betray a naive and short-term vision of humanity. Although it may not be fiscally realistic to prepare for war in the early twenty-first century, the extent to which postmodern societies are becoming dependent on private companies presents a security threat that needs to be addressed, and not just regulated, before it is too late.

Violence and control

Finally, the military use of non-state actors has challenged the legal control of the state and endangered the norms that limit the exercise of violence in warfare. The modern state has defined itself in terms of its ability to control the levels of violence within its territory. This has been contingent on the obedience of its soldiers who are held accountable for their actions. While the lawlessness of mercenaries has been an opportunity for states to decline responsibility and give a free reign to its unofficial representatives, the lack of legal control over private corporations and their employees effectively undermines the legitimacy of the hiring state and presents a threat to the security of the nation. Governments can no longer claim power on the basis of their ability to control the level of violence within their jurisdiction when they fail to hold their own agents accountable.

Furthermore, the absence of punitive measures for crimes committed by mercenaries encourages unlimited violence on the part of these actors. By failing to control the activities of their employees, the state is effectively sanctioning their crimes. This is particularly worrisome as it threatens to tear apart the international legal apparatus that limits the violence committed by military actors and protects the dignity of the victims of war. This can also increase the level of violence committed by all actors involved in the conflict, a horrifying perspective for anyone involved in humanitarian activities.

Mercenaries and contractors are an intriguing political tool, but they threaten to overhaul the entire military and social apparatus of the modern state in unexpected and potentially dangerous ways. This may change, as these non-state warriors increasingly become features of warfare, integrated into the conventional armed forces and regulated not only by voluntary codes of conduct but legally and criminally accountable to the state in which they serve.

Notes

1 "If you want peace, prepare for war" – Publius Flavius Vegetius Renatus's tract *De Re Militari.*
2 Bobbitt, *The Shield of Achilles: War, Peace, and the Course of History,* 819.

Bibliography

"A Look at US Deaths in the Iraq War". *Washington Post*, 25 October 2005.

"Blood Diamond". Dir. Edward Zwick, 2007.

"Contractors Accused of Firing on Civilians, Gis". *NBCNews*, 12 August 2007.

"Defence Spending Cuts Risk Military Skills, Warns Whitehall Watchdog". *BBC News*, 16 February 2012.

"De Imperatoribus Romanis on the Rulers of Rome: An Online Encyclopedia of Roman Rulers and Their Families".

"The Dogs of War". Dir. John Irvin, 1981.

"The Expendables". Dir. Sylvester Stallone, 2010.

"Fighting for Queen and Mother Country" www.blackinbritain.com/Feat%20070408%20 For%20Queen%20and%20Country.htm.

"The Fog and Dogs of War". *The Economist*. 2004.

"French 'Dog of War' Spared Jail". *BBC News*, 20 June 2006.

"German Soldiers Under-Equipped and Under-Trained, Says Politician". *German Radio*, 8 April 2010. www.defencetalk.com/german-soldiers-under-equipped-and-under-trained-says-politician-25549/.

"Iraq 'Death Squad Caught in Act'". *BBC News*, 16 February 2006.

"Pentagon: Some Explosives Possibly Destroyed". *NBC News*, 2004.

"Peter Fabricius: The Day of the Mercenary Coup Is Past". *Independent*, 27 August 2004.

"Private Security Industry Set to Double by 2010". *ANTARA News*, 16 September 2005.

"Putsch Aux Comores: Cinq Ans De Prison Requis Contre Bob Denard". *Le Monde*, 9 March 2006.

"Troops in Iraq 'under-Equipped'". *BBC News*, 10 August 2006.

"The Wild Geese". Dir. Andrew V. McLaglen, 1978.

A.D.M.A.F.L.E. "La Légion Etrangère De 1831 À Nos Jours" www.initiel.com/admafle/ spip.php?article1587.

Abrahamsen, Rita and Michael C. Williams. *Security Beyond the State: Private Security in International Politics*. Cambridge, UK; New York: Cambridge University Press, 2011.

AFP. "120 Civils Tués Par Miliciens Et Mercenaires Pro-Gbagbo". *La Libre Belgique*, 2011.

Alexandra, Andrew, Deane-Peter Baker and Marina Caparini. *Private Military and Security Companies: Ethics, Policies and Civilmilitary Relations*. London: Routledge, 2008.

Aristotle, W. D. Ross and Lesley Brown. *The Nicomachean Ethics*. Oxford; New York: Oxford University Press, 2009.

Arlinghaus, Bruce E. *Military Development in Africa: The Political and Economic Risks of Arms Transfers*, Westview Special Studies on Africa. Boulder, CO: Westview Press, 1984.

Arnold, Guy. *Mercenaries: The Scourge of the Third World*. New York: St. Martin's Press, 1999.

Aron, Raymond. *Peace & War: A Theory of International Relations*. New Brunswick, NJ: Transaction, 2003.

Avant, Deborah D. *The Market for Force: The Consequences of Privatizing Security*. Cambridge, UK; New York: Cambridge University Press, 2005.

Azan, Paul Jean Louis. *L'armeé D'afrique De 1830 à 185*2, Collection Du Centenaire De L'Algérie Archéologie Et Histoire. Paris: Plon, 1936.

Bailes, Alyson, Ulrich Schneckener and Herbert Wulf. "Revisiting the State Monopoly on the Legitimate Use of Force", edited by Geneva Centre for the Democratic Control of Armed Forces (DCAF), 2007.

Bailes, Alyson J. K., Keith Krause and Theodor H. Winkler. "The Shifting Face of Violence". Edited by Geneva Centre for the Democratic Control of Armed Forces (DCAF), 2006.

Baker, Deane-Peter. *Just Warriors, Inc.: The Ethics of Privatized Force*, Think Now. London; New York: Continuum, 2011.

Barkawi, Tarak. "State and Armed Forces in International Context", 2010.

Barlow, Eeben. "Eeben Barlow's Military and Security Blog", 2008.

Barlow, Eeben. *Executive Outcomes: Against All Odds*. Alberton, South Africa: Galago Books, 2007.

Barthorp, Michael. *Slogging over Africa: The Boer Wars 1815–1902*. Johannesburg: Jonathan Ball Publishers, 2002.

Bartov, Omer. "The Conduct of War: Soldiers and the Barbarization of Warfare". *The Journal of Modern History* 64, 1992.

BBC 4 Radio. "The Jacobite Rebellion" 2003.

Beck, Ulrich. *Risk Society: Towards a New Modernity*. London: Sage Publications, 1992.

Bell, David Avrom. *The First Total War: Napoleon's Europe and the Birth of Warfare as We Know It*. Boston: Houghton Mifflin Co., 2007.

Berman, Marshall. *All That Is Solid Melts into Air: The Experience of Modernity*. New York: Simon and Schuster, 1982.

Berman, Nicolas Florquin and Eric G. "Armed and Aimless: Armed Groups, Guns, and Human Security in the Ecowas Region", 2005.

Best, Geoffrey. *War and Society in Revolutionary Europe, 1770–1870*, Fontana History of European War and Society. Leicester, UK: Leicester University Press in association with Fontana Paperbacks, 1982.

The Berkley Center. "Ending Wars Well: The Aftermath of War – Reflections on Jus Post Bellum with Michael Walzer", 2010.

Beuret, Michel. "Quand Les Suisses Ensanglantaient L'Europe". *L'Histoire*, 2001, www2.unil.ch/unicom/allez_savoir/AS20/pdf_files/2.pdf.

Bismarck, Otto von. "Blood and Iron Speech". Prussian Parliament, 1862.

Blanning, T. C. W. *The French Revolution in Germany: Occupation and Resistance in the Rhineland, 1792–1802*. Oxford and New York: Clarendon Press; Oxford University Press, 1983.

Blanning, T. C. W. *The French Revolutionary Wars, 1787–1802*, Modern Wars. London; New York: Arnold; Distributed exclusively in the USA by St. Martin's Press, 1996.

Blanning, T. C. W. *The Pursuit of Glory: Europe, 1648–1815*. 1st American edn, The Penguin History of Europe. New York: Viking, 2007.

Bobbitt, Philip. *The Shield of Achilles: War, Peace, and the Course of History*. 1st Anchor Books edn. New York: Anchor Books, 2003.

Bobbitt, Philip. *Terror and Consent: The Wars for the Twenty-First Century*. 1st Anchor Books ed. New York: Anchor Books, 2009.

Bocca, Geoffrey. *La Légion! The French Foreign Legion and the Men Who Made It Glorious*. New York: Crowell, 1964.

Bogdanos, Matthew and William Patrick. *Thieves of Baghdad: One Marine's Passion for Ancient Civilizations and the Journey to Recover the World's Greatest Stolen Treasures*. 1st US edn. New York, NY: Bloomsbury Pub.: Distributed to the trade by Holtzbrinck Publishers, 2005.

Booth, Ken. *Statecraft and Security: The Cold War and Beyond*. Cambridge, UK; New York: Cambridge University Press, 1998.

Bourque, Nickand Jason Bicanic. "Shadow Company". Purpose Films, 2006.

Bremer, L. Paul. and Malcolm McConnell. *My Year in Iraq: The Struggle to Build a Future of Hope*. New York: Simon & Schuster, 2006.

Bruckberger, R. L. *One Sky to Share, the French and American Journals*. New York: P. J. Kenedy, 1952.

Bruneau, Thomas C. *Patriots for Profit: Contractors and the Military in US National Security*. Stanford, CA: Stanford University Press, 2011.

Bull, Hedley. *The Anarchical Society: A Study of Order in World Politics*. London: Macmillan, 1977.

Butcher, Tim. *Chasing the Devil: The Search for Africa's Fighting Spirit*. London: Chatto & Windus, 2010.

Buzan, Barry, Ole Wæver and Jaap De Wilde. *Security: A New Framework for Analysis*. Boulder, CO: Lynne Rienner Pub., 1998.

Caferro, William. *John Hawkwood: An English Mercenary in Fourteenth-Century Italy*. Baltimore: Johns Hopkins University Press, 2006.

Caputo, Philip. *A Rumor of War: With a Twentieth Anniversary Postscript by the Author*. 1st edn. New York: Henry Holt and Co., 1996.

Chapleau, Philippe. *Les Mercenaires; De L'antiquite a Nos Jours*. Rennes, France: Editions OUEST-FRANCE, 2006.

Chapleau, Philippe and François Misser. *Mercenaires S.A.* Paris: Desclée de Brouwer, 1998.

Chesterman, Simon and Chia Lehnardt. *From Mercenaries to Market: The Rise and Regulation of Private Military Companies*. Oxford: Oxford University Press, 2007.

Christakis, Nicholas A., and National Bureau of Economic Research. "An Empirical Model for Strategic Network Formation". In *NBER working paper series working paper 16039*. Cambridge, MA: National Bureau of Economic Research, 2010.

CIA. "World Factbook Angola", 2012.

Cilliers, Jakkie, Peggy Mason and Institute for Security Studies (South Africa). *Peace, Profit or Plunder?: The Privatisation of Security in War-Torn African Societies*. Halfway House: Institute for Security Studies, 1999.

Cilliers, Jakkie and Greg Mills. *Peacekeeping in Africa*. Halfway House, South Africa Braamfontein, South Africa: Institute for Defence Policy; South African Institute of International Affairs, 1995.

Cilliers, Mark and Jakkie Malan. "Mercenaries and Mischief: The Regulation of Foreign Military Assistance Bill *Institute for Security Studies* Ocasional Paper 25, 1997.

Citino, Robert Michael. *The German Way of War: From the Thirty Years' War to the Third Reich*, Modern War Studies. Lawrence, KS: University Press of Kansas, 2005.

Clapham, Christopher S. *Africa and the International System: The Politics of State Survival*. Cambridge: Cambridge University Press, 1996.

Clark, Ian. *Waging War: A Philosophical Introduction*. Oxford and New York: Clarendon Press; Oxford University Press, 1988.

Clausewitz, Carl von, Michael Howard, Peter Paret and Beatrice Heuser. *On War*, Oxford World's Classics. New York: Oxford University Press, 2006.

Coalition Provisional Authority. "CPA Order Number 17 Status of the Coalition, Foreign Liaison Missions, Their Personnel and Contractors". In *CPA/ORD/26*, edited by Coalition Provisional Authority, 2003.

Coalition Provisional Authority. "Registration Requirements for Private Security Companies". In *CPA/MEM/26*, edited by CPA, 2004.

"Code De La Défense", 2004.

Cockayne, James and Emily Speers Mears. *Beyond Market Forces: Regulating the Global Security Industry*. New York: International Peace Institute, 2009.

Cohen, Eliot A. *Citizens and Soldiers: The Dilemmas of Military Service*. Ithaca: Cornell University, 1990.

Coker, Christopher. *Barbarous Philosophers: Reflections on the Nature of War from Heraclitus to Heisenberg*. New York: Columbia University Press, 2010.

Coker, Christopher. *Ethics and War in the 21st Century*, LSE International Studies. London; New York: Routledge, 2008.

Coker, Christopher. *The Future of War: The Re-Enchantment of War in the Twenty-First Century*, Blackwell Manifestos. Malden, MA; Oxford, UK: Blackwell Pub., 2004.

Coker, Christopher. *Humane Warfare*. London; New York: Routledge, 2001.

Coker, Christopher. *The Warrior Ethos: Military Culture and the War on Terror*, Lse International Studies Series. London; New York: Routledge, 2007.

Colás, Alejandro, and Bryan Mabee. *Mercenaries, Pirates, Bandits and Empires: Private Violence in Historical Context*. New York: Columbia University Press, 2010.

Commission on Wartime Contracting. "Inattention to Contingency Contracting Leads to Massive Waste, Fraud, and Abuse", 2011.

Cooper, Adolphe Richard. *Born to Fight*. Edinburgh,: Blackwood, 1969.

Cooper, Stephen. *Sir John Hawkwood; Chivalry and the Art of War*. South Yorkshire: Pen & Sword Books Limited, 2008.

Cotton, Sarah, Ulrich Petersohn, Molly Dunigan, Q. Burkhart, Edward O'Connell Megan Zander-Cotugno and Michael Webber. "Hired Guns Views About Armed Contractors in Operation Iraqi Freedom". RAND, 2010.

Dailykos. "Private Financing of Private Military Companies". www.dailykos.com.

Debusmann, Bernd. "In Outsourced US Wars, Contractor Deaths Top 1,000". *Reuters*, 3 July 2007.

Diamond, Jared M. *Guns, Germs, and Steel: The Fates of Human Societies*. 1st edn. New York; London: W. W. Norton & Co. Jonathan Cape, 1997.

Dickinson, Laura. *Outsourcing War and Peace*. Yale University Press, 2011.

Dugdale-Pointon. "Mike Hoare (Congo Mercenary)". www.historyofwar.org/articles/people_hoare.html.

Duncan, Francis. *The English in Spain; or, the Story of the War of Succession between 1834–1840. Compiled from the Letters, Journals, and Reports of Generals W. Wylde, Sir Collingwood Dickson, W.H. Askwith; Colonels Lacy, Colquhoun, Michell, and*

Major Turner, R.A.; and Colonels Alderson, Du Plat, and Lynn, R.E., Commissioners with Queen Isabella's Armies. London: J. Murray, 1877.

Dunigan, Molly. *Victory for Hire: Private Security Companies' Impact on Military Effectiveness*. Stanford, CA: Stanford Security Studies, 2011.

Dunleavy, Patrick and Brendan O'Leary. *Theories of the State: The Politics of Liberal Democracy*. New York: Meredith, 1987.

Durant, Michael J. and Steven Hartov. *In the Company of Heroes*. New York: G. P. Putnam's Sons, 2003.

Eckholm, Erick. "US Contractor Found Guilty of $3 Million Fraud in Iraq". *The New York Times*, 10 March 2006.

Elliott, Lorraine M. and Graeme Cheeseman. *Forces for Good: Cosmopolitan Militaries in the Twenty-first Century*. Manchester: Manchester University Press, 2004.

Elsea, Jennifer, and Nina M. Serafino. *Private Security Contractors in Iraq Background, Legal Status, and Other Issues*. Washington, DC: Congressional Information Service, Library of Congress, 2004.

Elting, John Robert. *Swords around a Throne: Napoleon's Grande Armée*. New York: Free Press, 1988.

Englebert, Pierre. *Africa: Unity, Sovereignty, and Sorrow*. Boulder, CO: Lynne Rienner Publishers, 2009.

Evans, Peter B., Dietrich Rueschemeyer, Theda Skocpol, Social Science Research Council (US). Committee on States and Social Structures, Joint Committee on Latin American Studies., and Joint Committee on Western Europe. *Bringing the State Back In*. Cambridge, UK; New York: Cambridge University Press, 1985.

Fainaru, Steve. *Big Boy Rules: America's Mercenaries Fighting in Iraq*. Cambridge, MA; Philadelphia: Da Capo Press; Available from Perseus Books Group, 2008.

Fall, Bernard B. *Street without Joy*. Mechanicsburg, PA: Stackpole Books, 1994.

Feaver, Peter. *Armed Servants: Agency, Oversight, and Civil–Military Relations*. Cambridge, MA: Harvard University Press, 2003.

Feaver, Peter and Richard H. Kohn. *Soldiers and Civilians: The Civil–military Gap and American National Security*. Cambridge, MA: MIT, 2001.

Ferguson, Adam. *An Essay on the History of Civil Society*. 6th edn. 1 vols. n. p. 1793.

Figes, Orlando. *The Crimean War: A History*. 1st edn. New York: Metropolitan Books, 2010.

Finer, Samuel Edward. *The Man on Horesback; the Role of the Military in Politics*. New York: Praeger, 1962.

Force, Defense Science Board Task. "Report of the Defense Science Board Task Force on Outsourcing and Privatization". Washington, DC: Office of the Under Secretary of Defense for Acquisition and Technology, 1996.

Forsyth, Frederick. *The Biafra Story: The Making of an African Legend*. Barnsley, South Yorkshire: Pen & Sword Military, 2007.

French, Shannon E. *The Code of the Warrior: Exploring Warrior Values Past and Present*. Lanham, MD: Rowman & Littlefield Publishers, 2003.

Gat, Azar. *The Origins of Military Thought: From the Enlightenment to Clausewitz*. Oxford: Clarendon, 1989.

Gat, Azar. *War in Human Civilization*. Oxford; New York: Oxford University Press, 2006.

General Assembly. "56/232. Use of Mercenaries as a Means of Violating Human Rights and Impeding".

General Assembly. "Declaration on the Granting of Independence to Colonial Countries and Peoples". In *1514(XV)*, edited by United Nations, 1960.

"The Geneva Conventions of 1949 and Their Additional Protocols". Geneva, 2005.

George, Alexander L. and Andrew Bennett. *Case Studies and Theory Development in the Social Sciences*, BCSIA Studies in International Security. Cambridge, MA: MIT Press, 2005.

Geraghty, Tony. *Guns for Hire: The Inside Story of Freelance Soldiering*. London: Portrait, 2007.

Germani, Hans. *White Soldiers in Black Africa; Related from His Own Experiences*. Cape Town: Nasionale Beekhandel Beperk, 1967.

Gilbert, Adrian. *Voices of the Foreign Legion: The History of the World's Most Famous Fighting Corps*. New York: Skyhorse Pub., 2010.

Glastris, Charles and Paul Moskos. "Now Do You Believe We Need a Draft?" *The Washington Monthly*, 2001.

Glete, Jan. *War and the State in Early Modern Europe: Spain, the Dutch Republic, and Sweden as Fiscal-Military States, 1500–1660*, Warfare and History. London; New York: Routledge, 2002.

Government Accountability Office. "Rebuilding Iraq Actions Needed to Improve Use of Private Security Providers". US GAO, 2006.

Government Accountability Office. "Rebuilding Iraq Actions Needed to Improve Use of Private Security Providers". US GAO, 2005.

Government Accountability Office. "Securing, Stabilizing and Rebuilding Iraq Progress Report", 2008.

Graves, Robert. *Good-Bye to All That; an Autobiography*. London: J. Cape, 1929.

Grobbelaar, N., G. Mills and E. Sidiropoulos. *Angola: Prospects for Peace*. Johannesburg: SAIIA, 2003.

Grossman, Dave. *On Killing: The Psychological Cost of Learning to Kill in War and Society*. 1st edn. Boston: Little, Brown, 1995.

Guay, Louis. *Mouvements Sociaux Et Changements Institutionnels: L'action Collective à L'èRe De La Mondialisation*, Géographie Contemporaine. Sainte-Foy, Québec Saint-Nicolas: Presses de l'Université du Québec; Distribution de Livres Univers, 2005.

Hanson, Victor Davis. *Carnage and Culture: Landmark Battles in the Rise of Western Power*. 1st edn. New York: Doubleday, 2001.

Hart, Adrian Liddell. *Strange Company*. London: G. Weidenfeld & Nicolson, 1953.

Hedges, Chris. *War Is a Force That Gives Us Meaning*. 1st edn. New York: Public-Affairs, 2002.

Held, David. *Global Transformations: Politics, Economics and Culture*. Cambridge, UK: Polity Press, 1999.

Heller, Agnes and John F. Rundell. *Aesthetics and Modernity: Essays*. Lanham, MD: Lexington Books, 2011.

Hironaka, Ann. *Neverending Wars: The International Community, Weak States, and the Perpetuation of Civil War*. Cambridge, MA: Harvard University Press, 2005.

HMSO. "Report of the Committee of Privy Counsellors Appointed to Inquire into the Recruitment of Mercenaries (the 'Diplock Report')", 1976.

Hoare, Mike. *Congo Warriors*. Paladin reprint ed. Boulder, CO: Paladin Press, 2008.

Hobbes, Thomas and Michael Oakeshott. *Leviathan, or, the Matter, Forme and Power of a Commonwealth Ecclesiasticall and Civil*. New York: Touchstone, 2008.

Hodge, Katie. "London Olympics 2012: Troops Make up 50% of Park Security". *Independent*. N.p., 1 August 2012. www.independent.co.uk/news/uk/home-news/london-olympics-2012-troops-make-up-50-of-park-security-7998922.html.

Holmqvist, Caroline. "Private Security Companies: The Case for Regulation". *SIPRI Policy Paper Stockholm* 9, 2005.

Houngnikpo, Mathurin C. *Guarding the Guardians: Civil–Military Relations and Democratic Governance in Africa*. Burlington, VT: Ashgate, 2010.

Howard, Michael. *The Franco-Prussian War: The German Invasion of France, 1870–1871*. London; New York: Routledge, 2001.

Howard, Michael. *War in European History*. Updated edn. New York, NY: Oxford University Press, 2009.

Howe, Herbert M. *Ambiguous Order: Military Forces in African States*. Boulder, CO: Lynne Rienner Publishers, 2001.

Howe, Herbert. "Private Security Forces and African Stability: The Case of Executive Outcomes". *Journal of Modern African Studies* 36/2, 1998.

Hughes, Chris. "Self-Defeating: Mercenaries and Foreign Troops Could Fight for UK as 20,000 British Soldiers Face the Axe". *The Mirror*, 8 June 2012.

Human Rights First. "Human Rights First", 2008.

Human Rights Watch. "World Report", 1995.

Hummel, Pierre and Rebecca Englebert. "Let's Stick Together: Understanding Africa's Secessionist Deficit". *African Affairs* 104/416, 2005: 399–427.

Hunt, Lynn. *The Making of the West: Peoples and Cultures, a Concise History*. 4th edn. Boston, MA: Bedford/St. Martins, 2012.

Huntington, Samuel P. *The Common Defense; Strategic Programs in National Politics*. New York: Columbia University Press, 1961.

Huntington, Samuel P. *The Soldier and the State; the Theory and Politics of Civil-Military Relations*. Cambridge: Belknap Press of Harvard University Press, 1957.

icasualties.org. "Operation Iraqi Freedom", 2009.

IDEA. "Voter Turnout".

International Committee of the Red Cross. "Convention (III) Relative to the Treatment of Prisoners of War. Geneva, 12 August 1949", edited by International Committee of the Red Cross, 2005.

Isenberg, David. "Budgeting for Empire: The Effect of Iraq and Afghanistan on Military Forces, Budgetsm and Plans". Oakland, CA: The Independent Institute, 2007.

Isenberg, David. "Dogs of War: Blue on White". www.spacewar.com/reports/Dogs_of_War_Blue_on_white_999.html.

Isenberg, David. *Shadow Force: Private Security Contractors in Iraq*. Westport, CT: Praeger Security International, 2009.

Isenberg, David. "The Uncounted Contractor Casualties". *The Huffington Post*, 10 May 2011.

Jäger, Thomas and Gerhard Kümmel. *Private Military and Security Companies: Chances, Problems, Pitfalls and Prospects*. 1. Aufl. edn. Wiesbaden: VS Verlag für Sozialwissenschaften, 2007.

Jameson, Simon. "The French Foreign Legion: A Guidebook to Joining". Salvo Books, 2012.

Janowitz, Morris. *The Professional Soldier, a Social and Political Portrait*. Glencoe, IL: Free Press, 1960.

Jennings, Christian, and Copyright Paperback Collection (Library of Congress). *Mouthful of Rocks: Modern Adventures in the French Foreign Legion*. New York: Pocket Books, 1991.

Jocelyn, Ann. "Just How Overpaid Are Private Security Contractors?" *Serviam*, November/December 2007.

Josselin, Daphne and William Wallace. *Non-state Actors in World Politics*. Houndmills, Basingstoke, Hampshire: Palgrave, 2001.

Kaldor, Mary. *New & Old Wars*. Stanford, CA: Stanford University Press, 2007.

Katzenstein, Peter J. *Cultural Norms and National Security: Police and Military in Postwar Japan*, Cornell Studies in Political Economy. Ithaca, NY: Cornell University Press, 1996.

Keegan, John. *A History of Warfare*. New York: Alfred A. Knopf: distributed by Random House, Inc., 1993.

Keegan, John, Richard Holmes and John Gau. *Soldiers: A History of Men in Battle*. London: H. Hamilton, 1985.

Keen, David. *Conflict & Collusion in Sierra Leone*. Oxford, UK: James Currey, 2005.

Keen, Maurice Hugh. *Medieval Warfare: A History*. Oxford; New York: Oxford University Press, 1999.

Kemencei, Janos. *Légionnaire, En Avant!: De Budapest 1942, à Sidi-Bel-Abbès 1962*, Collection Vécu. Saint-Cloud: Atlante, 2000.

Kieh, George Klay. *Dependency and the Foreign Policy of a Small Power: The Liberian Case*. San Francisco, CA: Mellen Research University Press, 1992.

Kieh, George Klay, and Pita Ogaba Agbese. *The Military and Politics in Africa: From Engagement to Democratic and Constitutional Control*, Contemporary Perspectives on Developing Societies. Aldershot: Ashgate, 2004.

Kinsey, Christopher and Malcolm Hugh Patterson. *Contractors and War: The Transformation of US Expeditionary Operations*. Stanford, CA: Stanford Security Studies, 2012.

Kinsey, Christopher. "International Law and the Control of Mercenaries and Private Military Companies". *Cultures & Conflicts*, 2008.

Kinsey, Christopher. *Private Contractors and the Reconstruction of Iraq: Transforming Military Logistics*, Contemporary Security Studies. London; New York: Routledge, 2009.

Kopecki, Dawn. "Military Equipment: Missing In". *Bloomberg Businessweek*, 2007.

Korb, Lawrence J., Peter Ogden and Frederick W. Kagan. "Jets or GIs? How Best to Address the Military's Manpower Shortage". *Foreign Affairs*, 2006.

Krahmann, Elke. *States, Citizens and the Privatization of Security*. Cambridge, UK; New York: Cambridge University Press, 2010.

Krasner, Stephen D. *Problematic Sovereignty: Contested Rules and Political Possibilities*. New York: Columbia University Press, 2001.

Krasner, Stephen D. *Sovereignty: Organized Hypocrisy*. Princeton, NJ: Princeton University Press, 1999.

Le Mire, Henri. *Histoire Militaire De La Guerre D'Algérie*. Paris: A. Michel, 1982.

Leander, Anna. *Eroding State Authority? Private Military Companies and the Legitimate Use of Force*: Rubbettino Editore, 2006.

Légion-étrangère. "5. Le Règlement De La Légion Étrangère Est-Il Beaucoup Plus Dur Que Dans L'armée Française?" www.legion-recrute.com/fr/faq.php#f5.

Légion-étrangère. "Code D'honneur Du Légionnaire". Portail Internet de la Légion étrangère.

Légion-étrangère. "De 1946 À Nos Jours". Portail Internet de la Légion étrangère.

Légion-étrangère. "L'identité Déclarée". Portail Internet de la Légion étrangère, www.legion-etrangere.com/modules/info_seul.php?id=58.

Légion-étrangère. "La Carrière Dans La Légion Étrangère". www.legion-recrute.com/fr/carriere.php.

Légion-étrangère. "Salaires". www.legion-recrute.com/fr/salaires.php.

Lindström, Eric, James McPherson, Greystone Communications, History Channel (Television network), and Copyright Collection (Library of Congress). *The American Civil War*. United States: A&E Television Network New Video, 1994.

Lodge, Tom, Denis Kadima, David Pottie and Electoral Institute of Southern Africa. *Compendium of Elections in Southern Africa*. Johannesburg: Electoral Institute of Southern Africa, 2002.

Lords Spiritual and Temporal and Commons. "Bill of Rights". edited by Lords Spiritual and Temporal and Commons, 1689.

Lynn, John A. *Feeding Mars: Logistics in Western Warfare from the Middle Ages to the Present*. Boulder, CO: Westview, 1993.

National Security Archive Electronic Briefing. "Return of the Fallen", 2005, www.gwu.edu/~nsarchiv/NSAEBB/NSAEBB152/.

Louis-Philippe, Roi des Français. "Charte Constitutionnelle Du 14 Août 1830". Edited by Assemblée nationale, 1830.

Louis-Philippe, Roi des Français. "Loi Soult Du 21 Mars 1832". ANCESTRAMIL, www.ancestramil.fr/uploads/01_doc/organisation/loi_soult_1832.pdf.

Luttwak, Edward. *The Endangered American Dream: How to Stop the United States from Becoming a Third World Country and How to Win the Geo-Economic Struggle for Industrial Supremacy*. New York: Simon & Schuster, 1993.

Luttwak, Edward. "Where Are the Great Powers". *Foreign Affairs* 73/4, 1994: 23–29.

M6. "Legion Etrangere Documentaire" www.dailymotion.com/video/x4tplr_legion-etrangere-les-hommes-sans-no_travel.

Mabee, Bryan. *The Globalization of Security*. New York, NY: Palgrave Macmillan, 2009.

Machiavelli, Niccolò and N. H. Thompson. *The Prince*. Unabridged. edn, Dover Thrift Editions. New York: Dover Publications, 1992.

MacIntyre, Alasdair C., Paul Blackledge and Neil Davidson. *Alasdair Macintyre's Engagement with Marxism: Selected Writings 1953–1974*, Historical Materialism Book Series, Leiden; Boston: Brill, 2008.

Mancham, Sir James Richard Marie. "Mancham Sir James Richard Marie: Founding President of the Seychelles". http://jamesmancham.com/biography.php.

Manchester, William. *Goodbye, Darkness: A Memoir of the Pacific War*. 1st ed. Boston: Little, Brown, 1980.

Mandel, Robert. *Armies without States: The Privatization of Security*. Boulder, CO: Lynne Rienner, 2002.

Mann, Simon. *Cry Havoc: "When I Set out to Overthrow an African Tyrant, I Knew I Would Either Make Billions or End up Getting Shot*. London: John Blake, 2011.

Maogoto, Jackson Nyamuya. *State Sovereignty and International Criminal Law: Versailles to Rome*, International and Comparative Criminal Law Series. Ardsley, NY: Transnational Publishers, 2003.

Martin, Horace and Theodore. *The Odes of Horace*. London: J. W. Parker and Son, 1860.

Martinovic, Jovo. "Gaddafi's Fleeing Mercenaries Describe the Collapse of the Regime". *Time World*, 24 August 2011.

Martyn, Frederic. *Life in the Legion, from a Soldier's Point of View*. London: Everett & Co., 1911.

Mazrui, Ali Al Amin. *The Warrior Tradition in Modern Africa*, International Studies in Sociology and Social Anthropology. Leiden: Brill, 1977.

McNeill, William Hardy. *Plagues and Peoples*. Garden City, NY: Anchor, 1976.

McNeill, William Hardy. *The Pursuit of Power: Technology, Armed Force, and Society since A.D. 1000*. Chicago: University of Chicago, 1982.

McPhee, Peter. *The French Revolution, 1789–1799*. Oxford, UK; New York: Oxford University Press, 2002.

Mearsheimer, John. "The False Promise of International Institutions". *International Security* 19/3, 1994/5: 5–49.

Merchet, Jean-Dominique. "La Légion Étrangère S'accroche À Ses Effectifs". *Libération*, 6 November 2008.

Mills, Greg. *Why Africa Is Poor: And What Africans Can Do About It*. Johannesburg, South Africa: Penguin Books, 2010.

Mills, Greg and David Williams. *7 Battles That Shaped South Africa*. 1st edn. Cape Town: Tafelberg, 2006.

Mockler, Anthony. *Mercenaries*. London: Macdonald & Co., 1970.

Mollaret, Guillaume. "Ces Soldats Qui Désertent La Légion Étrangère". *Le Figaro*, 2007.

Moorcraft, Paul L. and Peter McLaughlin. *The Rhodesian War: A Military History*. Barnsley, England: Pen & Sword Military, 2008.

Moore, Darren. *The Soldier: A History of Courage, Sacrifice and Brotherhood*. London: Icon Books, 2009.

Morris, Donald Robert. *The Washing of the Spears: A History of the Rise of the Zulu Nation under Shaka and Its Fall in the Zulu War of 1879*. London: Cape, 1966.

Moskos, Charles. "Grave Decisions: When Americans Feel More at Ease Accepting the Casualties of War". *Chicago Tribune*, 12 December 1995.

Moskos, Charles C. *The American Enlisted Man; the Rank and File in Today's Military*, Publications of Russell Sage Foundation. New York: Russell Sage Foundation, 1970.

Moskos, Charles C., John Allen Williams and David R. Segal. *The Postmodern Military: Armed Forces after the Cold War*. New York: Oxford University Press, 2000.

Münkler, Herfried. *The New Wars*. Oxford: Polity, 2005.

Murray, Simon. *Legionnaire: An Englishman in the French Foreign Legion*. London: Sidgwick and Jackson, 1978.

Musah, Abdel-Fatau and Kayode Fayemi. *Mercenaries: An African Security Dilemma*. London; Sterling, VA: Pluto Press, 2000.

Myers, Lisa. "US Contractors in Iraq Allege Abuses". *NBC*, 17 February 2005.

National Security Archives. "Obama Administration Lifts Blanket Ban on Media Coverage of the Return of Fallen Soldiers". www.gwu.edu/~nsarchiv/news/20090226/index.htm.

North, Douglass Cecil. *Structure and Change in Economic History*. 1st ed. New York: Norton, 1981.

Norton-Taylor, Richard. "Limit on Commonwealth Troops Proposed to Keep Army 'British'". *Guardian*, 2 April 2007.

Norton-Taylor, Richard. "Mod May Halt Surge in Commonwealth Recruits to Army". *Guardian*, 5 April 2008.

O'Keefe, Padraig, and Ralph Riegel. *Hidden Soldier: An Irish Legionnaire's Wars from Bosnia to Iraq*. Dublin: O'Brien, 2007.

OAU. "OAU Charter". edited by Organisation of African Unity. Addis Ababa, 1963.

Olson, Mancur. *Power and Prosperity: Outgrowing Communist and Capitalist Dictatorships*. New York: Basic Books, 2000.

Ortiz, Carlos. *Private Armed Forces and Global Security: A Guide to the Issues*, Contemporary Military, Strategic, and Security Issues. Santa Barbara, CA: Praeger, 2010.

Owen, Wilfred. "Dulce Et Decorum Est", 1917.

Paret, Peter, Gordon Alexander Craig and Felix Gilbert. *Makers of Modern Strategy: From Machiavelli to the Nuclear Age*, Princeton Paperbacks. Princeton, NJ: Princeton University Press, 1986.

Pelton, Robert Young. *Licensed to Kill: Hired Guns in the War on Terror*. 1st edn. New York: Crown Publishers, 2006.

Pelton, Robert Young. *Licensed to Kill: Hired Guns in the War on Terror*. 1st pbk. edn. New York: Three Rivers Press, 2007.

Percy, Sarah V. *Mercenaries: The History of a Norm in International Relations*. Oxford; New York: Oxford University Press, 2007.

Perlmutter, Amos. *Political Roles and Military Rulers*. London; Totowa, NJ: F. Cass, 1981.

Perlmutter, Amos and Harvard University. Center for International Affairs. *The Military and Politics in Modern Times: On Professionals, Praetorians, and Revolutionary Soldiers*. New Haven: Yale University Press, 1977.

Perret, Antoine, *Montreux Five Years On: An analysis of state efforts to implement Montreux Document legal obligations and good practices*, American University Washington College of Law, 2013.

Peters, Ralph. "Blackwater vs. National Defense". *New York Post*, 6 October 2007.

Pilling, David. "The Trials of a Reluctant Superpower". *Financial Times*, 2 February 2012.

Plato and Benjamin Jowett. *The Republic*, Dover Thrift Editions. Mineola, NY: Dover Publications, 2000.

Poer, John Patrick Le. *A Modern Legionary*: Nabu Press 2010.

Polybius, Robin Waterfield and B. C. McGing. *The Histories*, Oxford World's Classics. Oxford: Oxford University Press, 2010.

Porch, Douglas. *The French Foreign Legion: A Complete History of the Legendary Fighting Force*. New York: Skyhorse Pub., 2010.

Posen, Barry R. *Inadvertent Escalation: Conventional War and Nuclear Risks*, Cornell Studies in Security Affairs. Ithaca, NY: Cornell University Press, 1991.

Posen, Barry R. "Nationalism, the Mass Army, and Military Power". *International Security* 18/2, 1993: 80–124.

Publius Flavius Vegetius Renatus. "De Re Militari".

Prado, José L. Gómez Del. "Private Military and Security Companies and the Un Working Group on the Use of Mercenaries". *Journal of Conflict & Security Law* 13/3, 2008: 429–450.

Radia, Kirit. "Controversial Blackwater Security Firm Gets Iraq Contract Extended by State Dept". *ABC*, 1 September 2009.

Rasmussen, Mikkel Vedby. *The Risk Society at War: Terror, Technology and Strategy in the Twenty-first Century*. Cambridge: Cambridge University Press, 2006.

Reichel, Clemens. "Catastrophe! The Looting and Destruction of Iraq's Past" The Oriental Institute of the University of Chicago, http://oi.uchicago.edu/museum/special/catastrophe/summary.html.

Remarque, Erich Maria, Helmuth Kiesel, and Joseph Roth. *All Quiet on the Western Front*, The German Library. New York: Continuum, 2004.

Richard, Jules. *La Jeune Armee*. Paris: La Librairie Illustree, 1890.

Ricks, Thomas E. *Fiasco: The American Military Adventure in Iraq*. New York: Penguin Press, 2006.

Rielly, Major Robert J. "Confronting the Tiger: Small Unit Cohesion in Battle". *Military Review*, 2000.

Ripley, Tim. *Mercenaries: Soldiers of Fortune*. Bristol: Parragon, 1997.

Roberts, Adam. *The Wonga Coup: Guns, Thugs, and a Ruthless Determination to Create Mayhem in an Oil-Rich Corner of Africa*. 1st edn. New York: PublicAffairs, 2006.

Roberts, Geoffrey. *Stalin's Wars: From World War to Cold War, 1939–1953*. New Haven CT; London: Yale University Press, 2006.

Rosen, Erwin. *In the Foreign Legion: The Experiences of a Journalist Who Joined the French Foreign Legion in North Africa at the Turn of the 20th Century* Leonaur Ltd 2010.

Rousseau, Jean-Jacques and Jean-Jacques Rousseau. *Discourse on Political Economy; and the Social Contract*, Oxford World's Classics. Oxford; New York: Oxford University Press, 2008.

Salazar, Jaime. *Legion of the Lost: The True Experience of an American in the French Foreign Legion*. New York: Berkley Caliber Book, 2005.

Sarat, Austin and Jennifer Louise Culbert. *States of Violence: War, Capital Punishment, and Letting Die*. Cambridge; New York: Cambridge University Press, 2009.

Saunders, Frances Stonor. *Hawkwood: Diabolical Englishman*: Faber and Faber, 2005.

Savych, Eric V. and Bogdan Larson. "American Public Support for US Military Operations from Mogadishu to Baghdad". RAND Arroyo Center, 2005.

Scahill, Jeremy. *Blackwater: The Rise of the World's Most Powerful Mercenary Army*. New York, NY: Nation Books, 2007.

Schultz, Sabrina. "The Good, the Bad and the Unregulated". *Institute for Security Studies*, 2008: 123–42.

Scott, George Ryley. *History of Torture*. London: Sphere, 1971.

Seeger, Alan and William Archer. *Poems*. New York: C. Scribner's Sons, 1917.

Segal, David R. *Recruiting for Uncle Sam: Citizenship and Military Manpower Policy*. Lawrence, KS: University of Kansas, 1989.

SÉNAT. "Proposition De Loi Relative À L'attribution De La Nationalité Française À L'étranger Qui a Combattu Dans Une Unité De L'armée Française". In *28*, 2000.

Shane, Leo III. "Report: US Wasted $60 Billion in Contracting Fraud, Abuse". *Stars and Stripes*, 31 August 2011.

Shanker, T. "Army Pushes a Sweeping Overhaul". *New York Times*, 4 August 2004.

Shaw, Martin. *Post-Military Society: Militarism, Demilitarization and War at the End of the Twentieth Century*: Polity Press, 1991.

Shay, Jonathan. *Achilles in Vietnam: Combat Trauma and the Undoing of Character*. New York Toronto: Atheneum; Maxwell Macmillan Canada; Maxwell Macmillan International, 1994.

Shearer, David. *Private Armies and Military Intervention*. Oxford: Oxford University Press for the International Institute for Strategic Studies, 1998.

Sheehy, Benedict, Jackson Nyamuya Maogoto and Virginia Newell. *Legal Control of the Private Military Corporation*. Houndmills, Basingstoke, Hampshire; New York: Palgrave MacMillan, 2008.

Simons, Suzanne. *Master of War: Blackwater USA's Erik Prince and the Business of War*. New York: Collins, 2009.

Singer, P. W. *Corporate Warriors: The Rise of the Privatized Military Industry*, Cornell Studies in Security Affairs. Ithaca: Cornell University Press, 2003.

Singer, P. W. *Corporate Warriors: The Rise of the Privatized Military Industry*. Updated ed, Cornell Studies in Security Affairs. Ithaca, NY: Cornell University Press, 2008.

Singer, P. W. "Can't Win with 'Em, Can't Go to War without 'Em: Private Military Contractors and Counterinsurgency". *Foreign Policy at Brookings* Policy Paper #4, 2007.

Singer, P. W. "Outsourcing War". *New York Times*, March 2005.

Singer, P. W. "Sure, He's Got Guns for Hire. But They're Just Not Worth It". *Washington Post*, 7 October 2007.

Slamdien, Fadela. "South Africa: Nation's Mercenary Legislation Remains Toothless". *West Cape News*, 2010.

Slater, Julia. "World Rulers from Switzerland". www.swissinfo.ch/eng/Home/Archive/World_rulers_from_Switzerland.html?cid=6817318, 2008.

Sledge, E. B. *With the Old Breed, at Peleliu and Okinawa*. New York: Oxford University Press, 1990.

Small Arms Survey. "Small Arms Survey 2007: Guns and the City", 2007.

Small, Michelle. "Privatisation of Security and Military Functions and the Demise of the Modern Nation-State in Africa". *The African Centre for the Constructive Resolution of Disputes (ACCORD)* Occasional Papers, 2006.

Smith, Sydney and Nowell C. Smith. *Selected Letters of Sydney Smith*, The World's Classics. Oxford; New York: Oxford University Press, 1981.

Social Science Research Council (US), and United States. Army Service Forces. Information and Education Division. *Studies in Social Psychology in World War II*. 4 vols. Princeton: Princeton University Press, 1949.

Spear, Joanna. "Market Forces: The Political Economy of Private Military Companies". In *New Security Report*: Fafo, 2006.

Spicer, Tim. *An Unorthodox Soldier: Peace and War and the Sandline Affair: An Autobiography*. Edinburgh: Mainstream, 1999.

Stanger, Allison. *One Nation under Contract: The Outsourcing of American Power and the Future of Foreign Policy*. New Haven: Yale University Press, 2009.

Stiglitz, Joseph E. *Globalization and Its Discontents*. 1st edn. New York: W. W. Norton, 2002.

Stouffer, Samuel Andrew. *The American Soldier*. 2 vols, Studies in Social Psychology in World War II, Princeton: Princeton University Press, 1949.

Stray, Jonathan. "What Did Private Security Contractors Do in Iraq?" Overview, http://overview.ap.org/blog/2012/02/iraq-security-contractors/.

Sun, Tzu Tzu. *The Art of War*. Middlesex: Echo Library, 2006.

Swan, Steven L. and Collin D. Schooner. "Dead Contractors: The Un-Examined Effect of Surrogates on the Public's Casualty Sensitivity". *Journal of National Security Law & Policy*, 2011.

Taylor, Ian. *The International Relations of Sub-Saharan Africa*. New York: Continuum, 2010.

Thomann, Jean-Claude. "La Réduction Des Effectifs Terrestres a Atteint Un Seuil". *Le Monde*. N.p., 16 July 2012. www.lemonde.fr/idees/article/2012/07/16/la-reduction-des-effectifs-terrestres-a-atteint-un-seuil_1734221_3232.html.

Thomas, Gerry S. *Mercenary Troops in Modern Africa*, A Westview Replica Edition. Boulder, CO: Westview Press, 1984.

Thomma, Steven. "Obama to Extend Iraq Withdrawal Timetable; 50,000 Troops to Stay". *McClatchy Newspapers*, 27 February 2009.

Thomson, Janice E. *Mercenaries, Pirates, and Sovereigns: State-Building and Extraterritorial Violence in Early Modern Europe*, Princeton Studies in International History and Politics. Princeton, NJ: Princeton University Press, 1994.

Thomson, Janice E. "The State, Sovereignty and International Violence: The Institutional and Normative Basis of State Control over External Violence". Thesis (Ph D), Stanford University, 1988.

Tilly, Charles. *Coercion, Capital, and European States, AD 990–1992*. Rev. pbk. edn, Studies in Social Discontinuity. Cambridge, MA: Blackwell, 1992.

Tilly, Charles. *The Politics of Collective Violence*, Cambridge Studies in Contentious Politics. Cambridge; New York: Cambridge University Press, 2003.

Tilly, Charles, Gabriel Ardant and Social Science Research Council. Committee on Comparative Politics. *The Formation of National States in Western Europe*, Studies in Political Development. Princeton, NJ: Princeton University Press, 1975.

Tomuschat, Christian. *Modern Law of Self-Determination*, Developments in International Law. Dordrecht; Boston: M. Nijhoff Publishers, 1993.

Tristam, Pierre. "Steve Fainaru's 'Big Boy Rules: America's Mercenaries Fighting in Iraq', 2008, http://middleeast.about.com/od/booksopinions/fr/me090105.htm.

UNHCR. "2012 Unhcr Country Operations Profile – Iraq". UNHCR 2012.

United Nations "The Exercise of the Right of Peoples to Self-Determination. In *A/RES/56/232*, 2002.

United Nations. "International Convention against the Recruitment, Use, Financing and Training of Mercenaries", edited by United Nations. New York, 1989.

United States Army. Judge Advocate General's Corps. *The Army Lawyer: A History of the Judge Advocate General's Corps, 1775–1975: With Finding Aids*. 1 vols. Buffalo, NY: W.S. Hein, 1993.

US Senate Armed Services Committee. *General Eric Shinseki*, 27 February 2003.

US Department of State. "Angola Country Specific Information". http://travel.state.gov/travel/cis_pa_tw/cis/cis_1096.html.

Van, Creveld Martin. *Technology and War: From 2000 B.C. to the Present*. New York: Free, 1989.

Van, Creveld Martin. *The Transformation of War*. New York: Free, 1991.

Vatican City. "History of the Swiss Guard". Vatican City, www.vatican.va/roman_curia/swiss_guard/swissguard/storia_en.htm.

Venter, Al J. *War Dog: Fighting Other People's Wars: The Modern Mercenary in Combat*. 1st ed. Philadelphia, PA: Casemate, 2006.

Venter, Al J. *War Stories: Up Close and Personal in Third World Conflicts*. 1st ed. Pretoria: Protea Book House, 2011.

Verkuil, Paul R. *Outsourcing Sovereignty: Why Privatization of Government Functions Threatens Democracy and What We Can Do about It*. New York: Cambridge University Press, 2007.

Wannenburg, Gail and South African Institute of International Affairs. *Africa's Pablos and Political Entrepreneurs: War, the State and Criminal Networks in West and Southern Africa*. Johannesburg: SAIIA, 2006.

Weber, Eugen. *Peasants into Frenchmen: The Modernization of Rural France, 1870–1914*. Stanford, CA: Stanford University Press, 1976.

Weber, Max, David S. Owen, Tracy B. Strong and Rodney Livingstone. *The Vocation Lectures*. Indianapolis: Hackett Pub., 2004.

Wellard, James Howard. *The French Foreign Legion*. 1st American edn. Boston: Little, Brown, 1974.

Wiking, Staffan. *Military Coups in Sub-Saharan Africa: How to Justify Illegal Assumptions of Power*. Uppsala Stockholm, Sweden: Scandinavian Institute of African Studies; Distributed by Almqvist & Wiksell International, 1983.

Wilson, John. *The History of Switzerland*. London: Printed for Longman, Brown, 1842.

Witney, Nick. "How to Stop the Demilitarisation of Europe". (2011), www.ecfr.eu/page/-/ECFR40_DEMILITARISATION_BRIEF_AW.pdf.

Wolvaardt, Pieter, Tom Wheeler and Werner Scholtz. *From Verwoerd to Mandela: South African Diplomats Remember*. 3 vols: Crink, 2010.

Wong, Leonard. "Why They Fight: Combat Motivation in the Iraq War", 2003. www.strategicstudiesinstitute.army.mil/pdffiles/pub179.pdf.

WorldPress. "U.S Soldier Dragged through Mogadishu". http://iconicphotos.wordpress.com/2010/03/10/u-s-marine-dragged-through-mogadishu/.

Wyatt, Thomas C. and Reuven Gal. *Legitimacy and Commitment in the Military*, Contributions in Military Studies, New York: Greenwood Press, 1990.

Yin, Robert K. "Case Study Research, Design and Methods". *Sage Publications*, 2003a: 13.

Young, Gary. "Don't Mention the Dead". *Guardian*, 7 November 2003.

Zarchin, Tomer. "International Legal Precedent: No Private Prisons in Israel". *Haaretz*, 20 November 2009.

Index

Page numbers in **bold** denote figures.